THE
AUDUBON SOCIETY
HANDBOOK
FOR
BIRDERS

Photo by Joe Van Os

THE AUDUBON SOCIETY HANDBOOK FOR BIRDERS

STEPHEN W. KRESS

Drawings by Anne Senechal Faust
Foreword by Olin Sewall Pettingill, Jr.

CHARLES SCRIBNER'S SONS · NEW YORK

In memory of Irving Kassoy,
who shared his enthusiasm
for watching birds

Copyright © 1981 Stephen W. Kress

Library of Congress Cataloging in Publication Data

Kress, Stephen W.
The Audubon Society handbook for birders.

Bibliography: p.
Includes index.
1. Bird watching.   I. National Audubon Society.
II. Title.
QL677.5.K73   598'.07'234   81-205
ISBN 0-684-16838-3   AACR2

1   3   5   7   9   11   13   15   17   19   V/C   20   18   16   14   12   10   8   6   4   2

Printed in the United States of America

All photographs by Stephen W. Kress, unless otherwise indicated.

CONTENTS

FOREWORD   ix
PREFACE   xi

1. FIELD TRIP TECHNIQUES   1

Identifying Birds   1
Shapes · Postures and Flight Patterns · Behaviors · Field Marks ·
Color · Size · Song · Habitat · Range and Abundance ·
Time of Year · Sorting Out Birds

Closing the Distance   13
Pishing and Squeaking · Mobbing Behavior · Playback Songs · Treetop Birding

Locating Birds   17
Leading a Bird Walk   20

2. BINOCULARS AND SPOTTING SCOPES   23

Selecting a Binocular   23
Power · Light-gathering Capacity · Field of View ·
Resolution · Alignment · Center Focus Versus Individual Focus
Binoculars · Roof Prism Binoculars · Mini Binoculars ·
Binoculars for Eyeglass Wearers · How to Shop for Binoculars

Focusing Binoculars   *33*
Binocular Care   *34*
How to Clean Binoculars · Protecting Binoculars
Selecting a Spotting Scope and Tripod   *35*
Birding with a Small Reflector Telescope · How to Shop
for a Bird-watching Telescope
Insuring Field Equipment   *38*

3. OBSERVING BIRDS   *39*
Watching Bird Behavior   *39*
Body Care Behaviors · Feeding Strategies · Social Displays
Behavior-watching Activities   *57*
Building an Ethogram · Identifying a Behavioral Sequence ·
Clocking a Time Budget · Publishing Your Observations
Taking Field Notes   *62*
Checklists · Field Notebooks · The Grinnell Field Note System ·
Note-taking Equipment · Preserving Your Field Notes · Punch Cards
Counting Birds   *70*
Sketching Birds   *73*
Assembling a Field Sketch

4. PHOTOGRAPHING AND RECORDING BIRDS   *83*
Selecting Photographic Equipment   *83*
Cameras · Telephoto Lenses · Flash Equipment
Film, Processing, and Home Printing   *90*
Identifying Photographic Problems   *90*
SPECtacular Photos · Angle of View · Distracting Backgrounds ·
"Soft" Images · Overexposures and Underexposures · Lighting Problems
Bird Blinds   *97*
Designs for Building a Portable Bird Blind
Bird Photographer's Etiquette   *105*
Bird Photography Programs   *106*
A Filing System for Bird Slides   *106*
Presenting a Slide Program   *109*
Resource Materials
Recording Bird Sounds   *113*
Selecting Equipment · Making Field Recordings

5. EDUCATIONAL PROGRAMS   *117*
North American Programs   *117*
Correspondence and Museum Programs · Field Programs
Ornithology Courses   *127*
International Bird Tours   *154*

6. RESEARCH PROGRAMS WELCOMING AMATEURS   *163*
    Breeding Season Programs   *165*
    Winter and Migration Programs   *174*
    All-season Programs   *178*
    U.S. State and Local Programs   *184*
    Canadian Provincial and Local Programs   *216*

7. PERIODICALS AND ORGANIZATIONS   *223*
    North American Periodicals and Organizations   *223*
    Professional Journals   *229*
    State and Provincial Periodicals   *231*
    State and Provincial Agencies   *241*

8. BUILDING A BIRD-WATCHER'S LIBRARY   *244*
    Bird Identification   *245*
    Life Histories   *249*
    Bird Biology   *253*
    Attracting Birds   *256*
    Bird-watching Activities   *259*
    Bird Clubs and Local Contacts   *260*
    Bird Finding   *261*
    North American and World Checklists   *263*
    Regional and Local References   *264*
    Birds of the World   *295*
    Bird Song Recordings   *297*

    APPENDIX A: Sources for Binoculars and Other Bird-watching Supplies   *302*
    APPENDIX B: Bird Book Retailers   *306*
    APPENDIX C: Bird and Bird-watching Publications from Natural Resource Agencies   *308*
    INDEX   *319*

THE NATIONAL AUDUBON SOCIETY

The National Audubon Society is among the oldest and largest private conservation organizations in the world. With 400,000-plus members and over 450 local chapters across the country, the Society works in behalf of our natural heritage through environmental education and conservation action and research. It protects wildlife in more than 70 sanctuaries from coast to coast. It also operates outdoor education centers, ecology workshops, and publishes the prize-winning *Audubon* magazine, *American Birds* magazine, newsletters, films, and other educational materials. For information regarding membership in the Society, write to the National Audubon Society, 950 Third Avenue, New York, New York 10022.

FOREWORD

Fifty years ago, after finishing college, I entered graduate school to obtain an advanced degree in ornithology and make ornithology my career. To my parents and friends the whole idea seemed dubious. How, they wondered, could I support myself studying birds? Why not enter medical school, as I once intended, and thus be assured of a livelihood in an established profession, and continue my interest in birds as an avocation?

That was in 1930. Ornithology, though little known as a profession, was nevertheless well established and thriving. Some twenty thousand books, pamphlets, and articles had been published. The American Ornithologists' Union, the leading professional society, consisted of two thousand members. About fifty colleges and universities were giving undergraduate courses in ornithology, and at least a dozen of the universities were granting advanced degrees in ornithology as a field of specialization. I could foresee opportunities for employment as a teacher. Furthermore, bird photography was coming into its own, offering opportunities for supplementary income. I was also encouraged by the fact that hundreds of people were supporting ornithology financially and contributing to the science in such sundry ways as undertaking life history studies, banding birds, and making

Christmas counts. Many people were on the move in their cars, over a steadily improving highway system, to explore hitherto remote parts of the country for birds. Guides to identification were in increasing demand, as were binoculars and telescopes.

But never in my wildest dreams did I imagine the extent to which ornithology would develop in the next fifty years or the tremendous following it would generate among people across the country. Today the volume of literature on birds is staggering. Guidebooks, handbooks, and textbooks, regional treatises, monographs on bird families or single species, and voluminous reports on behavioral and ecological studies now crowd library shelves; periodicals and newsletters flow from scores of bird-oriented societies and clubs. The American Ornithologists' Union has doubled its membership. More than six hundred academic institutions offer course work in ornithology, and there are many organizations—national, state, and local—that conduct programs or workshops to promote ornithological knowledge and/or to incite participation in significant projects such as censusing birds in designated areas.

Nonprofessional or amateur ornithologists, now generally called bird-watchers, number in the thousands. Many enjoy undertaking studies on their own; many others prefer joining in group projects. And there are thousands more people who are not bird-watchers in the studious or serious sense, yet like to photograph birds, to tape record bird songs, or to explore faraway places for viewing as many different kinds of birds as possible. No longer is any part of the globe too remote for bird-finding.

Bird-watching has become so vast a field for study and recreation, with so many ramifications, that a book is needed to point out the existing opportunities, describe the principal procedures and techniques as well as available equipment—binoculars, telescopes, cameras, and tape recorders—and provide references to essential information.

Stephen W. Kress has produced just such a book, a stimulating and instructive guide for anyone breaking into the bird-watching ranks and a *vade mecum* for all practicing bird-watchers.

Olin Sewall Pettingill, Jr.

PREFACE

This is an introductory technique manual and source book for one of the fastest-growing outdoor activities in North America. Widely known as bird-watching and fondly referred to as birding, it is a peculiar combination of a hobby, a sport, and a science. Because of this diversity and the appealing ways of birds, the popularity of this sport-hobby-science is rapidly gaining momentum.

Unlike many other outdoor activities that have a primary concern with equipment, the basic gear for bird-watching is simply binoculars, a notebook, and a field guide. Bird-watching's appeal is further broadened because it can be either a solitary or a social pursuit, and the pace of the activity will be as leisurely or energetic as one cares to make it.

Whatever the reasons for its appeal, bird-watching could become one of the most widely accepted outdoor pursuits in North America. Already more than half a billion dollars are spent each year for wild bird food, bird books, and binoculars, and each year more than three-quarters of a million bird field guides are sold. If the popularity continues to increase, there may be some very positive effects.

A recent study by the U.S. Fish and Wildlife Service reports that committed bird-watchers outshine other animal-oriented groups—such as zoo enthusiasts, pet owners, trappers, and supporters of animal welfare causes—by consistently demonstrating a "sophisticated and well-balanced environmental protection attitude." The study found that serious bird-watchers were the best-informed group on a broad range of ecological and animal life topics.

This research supports a widely held belief within the bird-watching community that an interest in birds is often a first step toward building a sound conservation ethic. Because birds are extremely sensitive indicators of the health of our environment, it is likely that a popular commitment to their well-being will lead to a more widespread concern for the quality of the environment on which we and the birds are mutually dependent. The conspicuous and appealing nature of birds makes them ideal candidates to help the American public understand our dependence on natural ecosystems and our responsibility to serve as stewards for wild birds and their varied habitats.

Approximately eighty-six hundred species of birds occur in the world, and more than seven hundred of these occur in North America. Keen observers can find more than one hundred different species moving through the vegetation and sky over most suburban yards in North America. For this reason, the best place to begin watching birds is at home. Start by identifying the birds that frequent your yard or nearby park, and do everything possible to improve your own property or local wildlife habitats. Take pride in the number of species you can encourage to nest or feed in your yard, but beware that bird identification and list keeping do not become ends in themselves.

Some veteran birders confide that bird-watching is not as much fun as it used to be. That feeling usually comes from listing without watching. To enjoy birds fully, keep observing after you've made identifications. The same excitement that comes from discovering "new" birds can be found by pursuing new activities and learning more about familiar birds.

This book is intended for both the beginning and experienced amateur bird-watcher. The beginner will find basic techniques such as how to identify birds and select binoculars and the right photographic equipment, while *both* the beginner and the experienced birder will find discussions of such field activities as sketching, note taking, photographic techniques, and sound recording. I have avoided reworking familiar topics such as bird attraction methods and listing favorite bird-watching locales because these topics receive thorough treatment elsewhere. To take the reader beyond the introductory material presented in this book, I include nearly five hundred annotated book recommendations and references.

Opportunities to participate actively in ongoing bird research and educational programs are overwhelmingly abundant. In an enthusiastic response to my call for research programs that welcome amateur participants, the professional ornithological community offered a total of 190 programs. Equally impressive are the over 600 ornithology courses and nearly 400 international bird-watching tours. Certainly there are many additional courses and research programs that do not appear on these lists, and the total represents only an indication of the opportunities that will be available to the amateur in the near future.

I could not have written this book without the enthusiastic support and encouragement of many people. For reading sections of the manuscript and offering suggestions for its improvement, I thank Robert M. Beck, Beth Campbell, William C. Dilger, Daniel Gray, Steven G. Herman, Michael J. Hopiak, Donald A. McCrimmon, Jr., and Alvin E. Staffan. I am also grateful to the following, who offered advice and assistance in assembling material: Kathleen A. Anderson, Robert S. Arbib, Jr., Diane L. Deluca, Erica H. Dunn, Thomas L. Fleischner, James L. Gulledge, David Klinger, Douglas A. Lancaster, Ronald Renault, Chandler S. Robbins, and Charles R. Smith.

I am also especially grateful to James A. Tucker and Susan Roney Drennan for their generous permission to reprint their annotated comments about regional bird references. It is a special pleasure to acknowledge the imaginative art of my illustrator, Anne Senechal Faust. Her work adds greatly to the value of the book. I also thank Beth Campbell for her very capable assistance in typing the manuscript, and I thank Leslie McKim for preparing many of the photographs.

Finally, my sincere thanks go to all those who took the effort to respond to my questionnaires, advertisements, and correspondence for assembling the directories to research and educational programs. The substantial nature of these listings is a testimony to their enthusiastic response.

Throughout my preparation of the manuscript, I found ready support for my principal purpose in writing the book—to provide a guide that would demonstrate that bird-watching can be a life-long pursuit that becomes more enjoyable as the years go by. I hope the book accomplishes its purpose, for I believe it will be a happier and gentler world if more people take the time to watch the birds.

STEPHEN W. KRESS
Ithaca, New York

1

FIELD TRIP TECHNIQUES

IDENTIFYING BIRDS

The ability to identify birds is largely a matter of familiarity that soon comes to anyone who concentrates on the differences and similarities among the birds that surround us daily.

Identifying birds is much like getting to know human neighbors. Move into a new community and at first everyone is a stranger. The ability to distinguish new personalities, faces, and voices is simply a matter of repeated contact and attention to detail. Soon, neighbors can be recognized on the street, and even a word over the telephone is enough to identify the caller. Look for the characteristics of the birds that visit your yard and you'll start to notice both the similarities and the differences.

Professional ornithologists group birds into families that contain closely related species. Families such as warblers, vireos, flycatchers, and finches are each comprised of species that show similarities to one another. If you learn to recognize birds by family characteristics, it becomes easy to identify the birds you see. Here are some things to look for.

Shapes

Take a close look at birds and you'll soon note many different shapes. Birds of approximately the same size vary dramatically in body shape and proportion. Some, like doves and pigeons, have chunky bodies, while others, such as mockingbirds and thrashers, have slender ones. Shapes of beaks, tails, wings, legs, and necks often vary markedly from one kind of bird to another, but these features are usually similar for birds within the same family. Knowledge of bird shapes (see illustration 1) will permit you to place birds in their families and recognize many by their characteristic silhouettes.

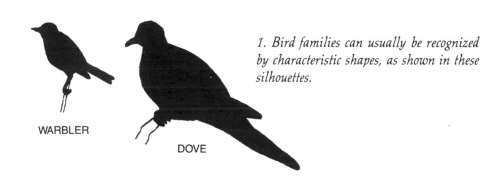

1. Bird families can usually be recognized by characteristic shapes, as shown in these silhouettes.

WARBLER

DOVE

WOODPECKER

Postures and Flight Patterns

You can usually place birds in the proper family by looking for similarities in posture. Watch a robin strut across the yard. Several steps forward are followed by an alert, upright stance. Other members of the thrush family have similar postures. Flycatchers perch upright, while vireos usually perch in a more horizontal posture (see illustration 2).

Birds also have characteristic flight patterns (see illustration 3). Note that finches have a steep, roller-coaster flight pattern, while woodpeckers characteristically fly in a deep, undulating manner. Even the way a bird holds its wings while in flight may be useful in making an identification at the family level. Vultures, for example, are easily distinguished from hawks and eagles by the position of their wings as they soar; vultures hold their wings at an upward angle over their backs, while hawks and eagles hold their wings on a level plane.

Behaviors

Warblers and vireos look very similar. Both families contain small, insect-eating birds that live high in trees. If you have the opportunity to observe them feeding, it will soon become apparent that warblers are quick, energetic feeders that seldom stay in one place as they pick tiny insects from leaves and branches. In con-

LEAST

GREAT-CRESTED

2. *These silhouettes of three flycatcher species show the similar vertical postures among the species. In contrast, vireos* (at bottom) *usually perch in a horizontal posture.*

PEWEE

RED-EYED VIREO

3. *Birds often show family similarities in their flight patterns—finches characteristically show a roller-coaster flight path, while woodpeckers exhibit an undulating flight pattern. The flight silhouettes of the hawk and vulture show the position of the wings as they soar.*

FINCH

WOODPECKER

HAWK

VULTURE

4. After you have identified a bird's family, watch it carefully for characteristic behaviors. For example, the phoebe is the only flycatcher to wag its tail.

trast, vireos often perch for several minutes in one place waiting until they see a large insect—then they dash forward to snatch up their prey.

Birds often have unique behaviors that distinguish them from closely related species in the same family. Such behaviors help birds recognize other members of their own species. The same behaviors can also help you make bird identifications. If you spot a slim bird sitting in an upright posture on a dead branch, it might be a member of the flycatcher family. If you observe it fly out, snatch an insect out of midair, and then return to the same perch, it probably *is* a flycatcher, since few other birds exhibit this "hawking" behavior. If the bird then starts wagging its tail, you can be almost certain that it is a phoebe—a medium-sized flycatcher that characteristically flicks its tail up and down (see illustration 4). Since there are three kinds of phoebes in North America, you would then have to refer to a field guide to find the differences in range and specific field marks.

Field Marks

The same distinctive patterns and colors that make birds so appealing to humans often serve as species identification signals for the birds themselves. The bold markings on male ducks, for example, usually ensure that the females have no difficulty in finding a male of the proper species. Conspicuous color patterns reduce the chance of interbreeding and the production of hybrids, which are often infertile or poorly adapted.

Field marks are also a useful aid to bird-watchers. Even inconspicuous markings may be useful to both birds and bird-watchers when it comes to distinguishing similar-looking species. Look the bird over for such markings as wing bars, eye rings, eye lines, eye stripes, crests, wing patches, and tail spots. Review the parts of a bird for quick reference to field marks (see illustration 5).

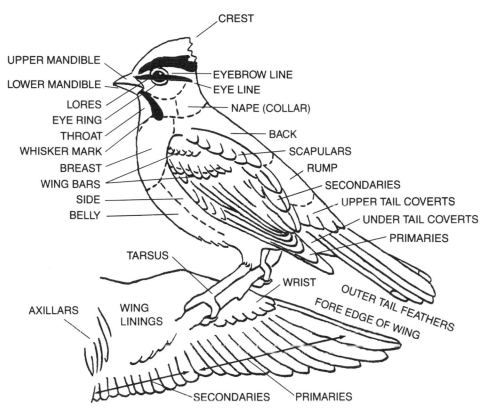

CREST
UPPER MANDIBLE
LOWER MANDIBLE
LORES
EYE RING
THROAT
WHISKER MARK
BREAST
WING BARS
SIDE
BELLY
EYEBROW LINE
EYE LINE
NAPE (COLLAR)
BACK
SCAPULARS
RUMP
SECONDARIES
UPPER TAIL COVERTS
UNDER TAIL COVERTS
PRIMARIES
TARSUS
WRIST
OUTER TAIL FEATHERS
FORE EDGE OF WING
AXILLARS
WING LININGS
SECONDARIES
PRIMARIES

5. Memorize the field marks and regions of this generalized bird so that you can use standard terminology when describing unknown birds.

Color

Although color can sometimes be a useful aid in bird identification, it is usually much better to rely on such clues as shape, posture, and behavior, since the apparent color of a bird often varies with the angle of view and lighting situation. Indigo buntings are one of the most dramatic examples of this. In direct light this bird flashes a brilliant indigo glow, but if the bird is lit from behind, it appears to be jet black in color. To demonstrate this effect, reflect light from a small pocket mirror onto a backlit indigo bunting. Suddenly the bird will glow as if a blue light has been turned on!

Another problem with learning to identify birds by color is that individual birds often vary in the amount of color they display, depending on age, sex, and the time of year. Birds such as tanagers, warblers, and shorebirds experience complete body molts during which all of their bright breeding-color feathers are lost as they change into more somber winter plumages. Also remember that it is confusing to make exact comparisons of living birds with color plates in field guides, since color reproductions only approximate bird colors, and the quality of the color varies from one printing to the next.

Patterns of contrasting color are more useful than actual colors. Patterns such as striped breasts, dark caps, and light-colored rumps are visible under most lighting conditions and are often conspicuous even at a considerable distance.

Size

Size is one of the most difficult features to use in bird identification. Like color, the apparent size of a bird is often distorted by lighting conditions and distance. Size can be especially confusing at dusk and dawn and under conditions of fog and rain. Birds can also change their apparent size depending on the weather. When the temperature is hot, birds can sleek their feathers tight to their bodies, which makes them look smaller. On frigid winter days they may fluff themselves up, providing better insulation and a much larger appearance.

6. The common crow, American robin, and house sparrow are common birds that can serve as size references for comparison with less familiar birds.

Size is most useful for making comparisons. The common crow, American rob-in, and house sparrow (see illustration 6) are common birds throughout most of North America and can serve as size references for comparison with less familiar birds. Comparisons are most useful if you see an unknown bird in company with one of the references, but with some practice the references can be reliable from memory—a raspberry-colored bird the size of a house sparrow may be a purple finch. Without a careful size comparison, it could be confused with a pine gros-beak, which has similar colors and patterns but is almost as big as a robin.

A woodpecker the size of a crow is no doubt a pileated woodpecker, while a woodpecker the size of a sparrow is probably a downy woodpecker. If an un-known bird is larger or smaller than a reference bird, then use two reference birds for comparison. Screech owls are larger than sparrows, but only slightly smaller than robins. Great horned owls are larger than crows.

Although birds of the same species are usually similar in size, you can some-times note a size difference between males and females. For example, male gulls, turkeys, and pheasants are larger than females of the same species, but female hawks, eagles, and owls are larger than males.

With few exceptions, young birds are as large as or larger than their parents when they leave the nest (see illustration 7). Exercising soon trims them to adult proportions. If you see a bird flying, assume that it has reached full adult size.

7. *When young birds, such as this fledgling grackle, leave their nest, they are nearly as large as their parents.*

Song

The primary function of bird song is to announce the presence of the singer by broadcasting the territorial message. Male songbirds defend their territories pri-marily through singing displays. They usually back their songs up with aggres-sive chases only if their territories are invaded by a rival male of the same species.

The songs and calls of birds are useful to bird-watchers because they often announce the identity of birds that would be otherwise overlooked in dense vegetation. Familiarity with bird songs usually comes after you have learned to recognize birds by their appearance. Like identifying birds by sight, song recognition comes only after much practice in the field.

Tape recordings and phonograph records of bird songs do not take the place of field experience but are most useful for review of songs heard outdoors. The best way to learn bird song is to concentrate on one song at a time and pursue the unknown singer until you get a good enough view to establish its identity.

Habitat

Birds are predictably associated with specific habitats. American bitterns, for example, are usually found in cattail marshes and usually live in association with other freshwater marsh birds such as swamp sparrows and king rails. In a salt marsh habitat, you are less likely to find a bittern (see illustration 8). The swamp sparrow would be replaced by a sharp-tailed sparrow or seaside sparrow, and clapper rails would replace king rails. Each plant community, such as a spruce–fir forest, a meadow, or a freshwater marsh, has a predictable complement of birds, which is part of the community. Learn which birds to expect in each habitat

8. The American bittern is easily camouflaged in its usual marshland habitat and seldom visits other habitats. When it is threatened by a predator, its brown-striped plumage and its habit of freezing in a vertical stance keep it hidden.

9. During migration, tired bitterns occasionally land in habitats where their specialized camouflage and behavior are of little value, and they become especially conspicuous.

and you'll be able to eliminate many similar species that are usually associated with other habitats.

Because birds often migrate over a variety of habitats between their breeding and wintering homes, they sometimes show up in surprising places. During spring and fall migrations, birds often settle down when they become tired, regardless of the habitat. Perhaps mistaking them for rivers, loons and ducks sometimes settle down on wet highways in error. Tired bitterns sometimes land in grassy backyards, but being so accustomed to hiding among cattails, they still hold their heads in a characteristic vertical posture (see illustration 9).

Knowledge of habitat is one of the best ways to identify similar-looking birds such as the alder, least, and Acadian flycatchers (see illustration 10). These small flycatchers all have white eye rings and wing bars. They are so similar in appearance that you could not easily tell them apart, even if you held them in your hand. The birds probably recognize members of their own species by their distinctive songs and preferences for different habitats. Alder flycatchers live in wet, marshy habitats; least flycatchers live in dry, scrubby fields; Acadian flycatchers live in moist woodlands.

ALDER

LEAST

ACADIAN

10. One way to identify similar-looking birds, such as the alder, least, and Acadian fly-catchers shown here, is by the habitat in which they are observed.

Range and Abundance

Although birds are highly mobile and often show up in out-of-the-way places, they are usually very predictable about staying within defined geographic limits known as ranges. General information about bird distribution is readily available in any of the North American bird field guides. Those with maps are the easiest to use in the field. Maps in some guides also provide information about breeding and wintering ranges and dates of arrival for migratory birds.

Range maps are invaluable aids in determining which of several similar species

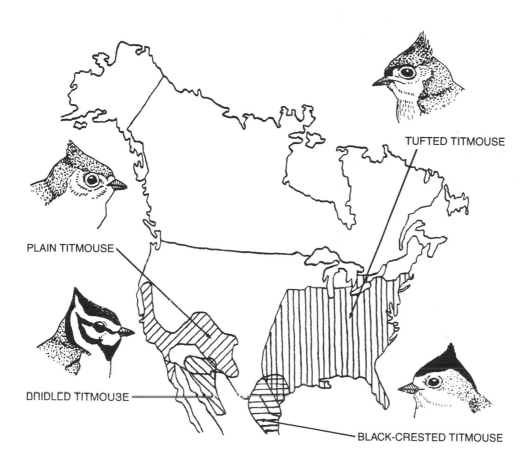

TUFTED TITMOUSE

PLAIN TITMOUSE

BRIDLED TITMOUSE

BLACK-CRESTED TITMOUSE

11. Bird range maps are one of the most important tools in establishing which of several common species you are most likely to see. For example, the ranges shown on this map for four species of titmouse do not usually overlap and are useful in showing which species one might expect to see.

you might hope to find in your region. Such maps can also show that for some species groups, such as titmice (see illustration 11), there is little range overlap. Just as it is important to know which birds you might see, it is also helpful to know something of the relative abundance of the different species. Learn the common birds first, then you will be more likely to spot birds that look different. Remember, however, that the birds have not seen the range maps or read the books. Ranges change and wandering individuals occur in most species. Your alert observations can help to document such events.

TREE SPARROW

FIELD SPARROW

12. The time of year is also important to consider when identifying similar-looking species. The ranges of the tree sparrow and field sparrow overlap, but the tree sparrow occurs in winter, while the field sparrow occurs after the tree sparrow has headed back north.

Time of Year

It is useful to know which birds visit your bird-watching locale at different times of the year. Similar-looking species can sometimes be told apart by knowing which birds to expect at each season. Some closely related species neatly divide the year with little overlap. Tree and field sparrows are two such examples (see illustration 12). Tree sparrows breed in the northern tundra and invade southern Canada and the United States only during the winter. About the time the last of the tree sparrows head back north, the field sparrows arrive from their winter habitat in the southern United States. Both species are seldom seen at the same time. Specific field marks, such as the tree sparrow's black chest spot and the field sparrow's pink bill, can confirm your observations.

Sorting Out Birds

The process of bird identification begins when you look for the features of shape, size, posture, and behavior that permit placing a bird in the correct family. With the family in mind, the number of possibilities is greatly reduced, and you need only next consider which members of the family are likely to be in this specific habitat at this time of year. Listen for song and look for field marks to make your final decision as to which species you are watching.

The ability to make quick field identifications comes slowly at first, but by looking for similarities within bird families, it will become increasingly easy to identify new members of the families you already know. Knowledge of bird families provides the great advantage of permitting quick identification of new species in distant places, as these will remind you of familiar birds. You'll find birds wherever you go, but the best place to begin watching them is at home.

CLOSING THE DISTANCE

The greatest problem for beginning bird-watchers is the distance that most birds prefer to keep between themselves and observers. Since most birds are usually wary of close approaches and quickly take flight or retreat into dense vegetation, it is useful to have a few tricks to help close the distance between you and the birds.

Pishing and Squeaking

To attract land birds try imitating the generalized distress call, known as pishing, that many birds give when they are alarmed. This call sounds like the word "pish" pronounced as a drawn-out, hissing exhale. It is most effective during the breeding season when birds are protecting their nesting territories. Then they are especially curious about unusual sounds and frequently approach to chase away predators. The call works best when you hear birds giving chip calls. Chip calls are short, kiss-like sounds produced when a bird is already upset by your presence or some other disturbance. Pishing may then bring the bird into view for closer observation.

Squeaking is a similar attraction technique. Try this by kissing the back of your hand to produce a prolonged squeaking sound or purchase the commercially made bird squeaker called the Audubon Bird Call. By twisting the metal core of the birdcall with varying degrees of pressure against the wooden exterior, the call will produce a variety of realistic chirps, squeaks, and trills (see illustration 13).

13. Squeaking (left) *is one attraction technique, in which the birder kisses the hand as shown. The Audubon Bird Call* (lower right) *can produce a variety of realistic chips and squeaks.*

Not all birds are equally attracted by the pishing and squeaking techniques. Much depends on the bird's breeding condition and level of excitement when you first attempt to attract it. Some birds, such as chickadees, will give their own alarm calls if they become excited by your pishing and squeaking efforts, and these calls will help attract additional birds.

14. Small birds often mob a predator in an effort to chase it away. Here red-winged blackbirds mob a great horned owl.

Mobbing Behavior

Songbirds are probably attracted to pishing and squeaking noises because these sounds are similar to the alarm notes that communicate the presence of a predator. Rather than fly away from threatening predators, small birds will often swoop and dash at a predator in an effort to make it leave their territory. Mobbing behavior is commonly expressed toward owls, hawks, and snakes. Even mammal predators, such as foxes, cats, or squirrels, are usually attacked by mobs of small birds if they are discovered near active bird nests.

Owls spark the most intense mobbing behaviors (see illustration 14). Although owls can see as well as most birds in daylight, they seldom attack small birds during the day. At night, however, some owls, especially screech and great horned owls, frequently prey upon sleeping birds. Small birds that chase owls out of their territories during daylight will be safer at night.

You can use the mobbing-behavior response to lure birds in for a close view. Just play a recording of a screech owl call and watch the reaction. The screech owl's trembling whistle usually attracts small birds, which flock to the sound ready to mob the owl. Even a whistle imitation of the screech owl's call will often rally a congregation of local songbirds. It's interesting to observe how all the different species that share territories will join in a combined effort to mob a common threat.

You can trigger an even stronger mobbing reaction by playing a screech owl recording in the presence of a papier-mâché owl model. Construct an owl by first building a chicken wire frame. Then wrap paper strips dipped in a mixture of

15. *Mobbing reaction can be triggered by placing a papier-mâché owl painted in realistic colors in a conspicuous place.*

flour and water over the chicken wire (see illustration 15). Paint the owl with realistic colors, being sure to give it big yellow eyes. Mount the owl in a conspicuous place, turn on the tape recorder, and hide nearby to watch the reaction.

Playback Songs

Bird song alerts males of the same species that a breeding territory is occupied. If a newcomer sings within an established territory, he will soon encounter the established male. Persistent song from the challenging male is usually followed by a chase from the territorial bird.

This chase response can be used to lure seldom-seen birds into view. Play a tape-recorded song within a bird's defended area, and the territorial male will be quick to put in an appearance. Seldom-seen birds that live in the densest tangles and treetops will come into view to challenge the newcomer in their territory.

Tape-recorded songs may be made either by copying available recordings or by recording the voice of the same territorial male you are trying to observe. Because birds cannot recognize their own voices, a singing male will come forward to defend his territory against any rival, even if the "rival" is his own recorded voice.

While an occasional confrontation with a tape recorder probably has little effect on a breeding bird's territorial behavior, care should be taken not to use recordings in excess. Once the bird you are seeking has appeared, turn off the recorder and let the male sing his song without competition. Never use tape recordings to lure into view rare or endangered birds or any bird that is nesting outside its normal range. Such a disturbance to a bird already in precarious circumstances may threaten breeding success.

Treetop Birding

Not all birds are shy of people. Many land birds, such as warblers, vireos, and kinglets, will often approach surprisingly close if you meet the birds at their own level. Birds that live in treetops often are not afraid of people who climb out on a branch to meet them. Find a comfortable perch and enjoy the view.

LOCATING BIRDS

Probably the greatest frustration met by beginning birders is the difficulty of finding birds with binoculars. Spotting birds with the unaided eye is one thing, but it's a very different challenge to find a bird in the narrow field of a binocular.

The most important tip for finding birds with binoculars is to first spot the bird with your eyes. Then, holding your head rigidly fixed to the bird, lift the binoculars to your eyes. Avoid wildly searching the trees with your binoculars. Practice locating stationary objects first, such as birdhouses, feeders, flowers, and tree branches. Locate large objects with your binoculars, then smaller forms. Practice spotting familiar birds that frequent feeders and live in open places before looking for birds in dense vegetation.

It's difficult to remain patient when you're the only one in a group who can't find a bird. Remember that your best chance of spotting a bird is with your unaided eyes. Then lift your binoculars.

If you are with a group of birders and cannot find a bird that the others see, ask for specific directions for locating the bird. Vague directions are no help and only increase the chance that the bird will fly away before you see it.

Always be specific when describing bird locations. Refer to the most obvious landmark near the bird, and narrow the field until you come to the bird. For example, if you spot a hawk in a farmland situation, you might describe its location to a fellow birder as follows: "Do you see that large red barn with the white silo? Look over the top of the silo to the fence behind. Are you with me? Count eight fence posts to the right and there's the hawk sitting on top of the post! Do you see it?" Frequent checks to make certain your directions are clear will increase your success in assisting others to spot birds efficiently. At moments of excitement, it takes as much skill to share your bird discovery as it does to locate birds in the first place.

If a bird is in a tree, use the "clock" technique to describe its position. Mentally superimpose an hour hand onto the tree (see illustration 16) and use it to point to the bird. If the bird is at the edge of the tree, the system works fairly well. A bird in the top of the tree is obviously at twelve o'clock; halfway down the right side it is at three o'clock. If the bird is not at the edge, then the hour designation is only the first step in describing the bird's position. In addition to the hour description, one must give other pointers, such as: "Find two o'clock in the largest sycamore, then move in halfway to the center of the tree. The bird is in front of

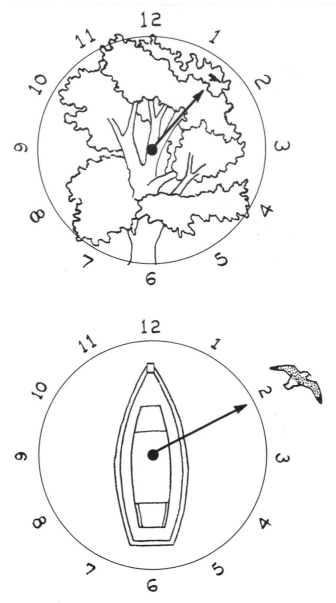

16. The clock technique is especially useful in describing bird locations to other birders. This diagram shows the "clock" and an hour hand superimposed on a tree and a boat.

the largest branch near a large woodpecker hole. See it?" Avoid using distance measurements such as "20 feet from the top of the tree." Most people find it difficult to agree on exact distances.

The clock system also works for spotting birds from a moving vehicle, such as a bus or a boat. For nautical birding, the clock is viewed as it would be seen if you were above the boat looking down, with the boat in the center of the clock (see illustration 16). Saying "A Ross's gull at one o'clock!" would certainly send people rushing forward to locate the bird just to the right of the bow. Similarly, the clock can be superimposed on land in a horizontal position. For a land-based clock, twelve o'clock usually points north, or toward some predetermined landmark. This system is frequently used to point out migratory hawks (see illustration 17).

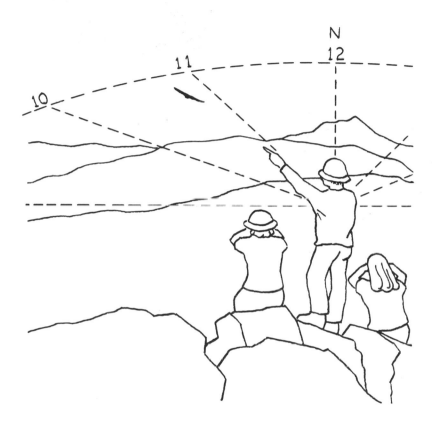

17. The clock technique can also be used on land in a horizontal position.

If the sun is out, a pocket mirror can be used to point out the location of birds by flashing a beam of light as a pointer. This is especially useful for pointing out nests that are high in trees or hidden in dense foliage.

The buddy binocular is a novel but compact and inexpensive way to point at birds (see illustration 18). Instead of sighting down an outstretched arm, the buddy binocular will focus a friend's attention on the exact spot where a bird is perched. A buddy binocular is constructed by cementing with silicon bathtub

18. The buddy binocular helps to focus another birder's attention on the exact spot where a bird is perched.

sealer two 5-inch sections of ¾-inch PVC or plastic plumbing pipe to a wooden block about ¾ inch by 5 inches by 3 inches.

LEADING A BIRD WALK

Sharing time in the field with family and friends is one of the greatest satisfactions arising from an interest in birds. A few enjoyable experiences observing birds can spark a lifetime interest, but the challenge comes in trying to make bird encounters meaningful to the novice who has little previous interest. You don't have to be an "authority" to share the excitement of identifying and watching birds. Anyone can communicate enthusiasm if he or she follows a few common-sense tips. Here are some suggestions to keep in mind:

1. *Group Size.* The optimal size for a bird-walk group is twelve or fewer. A group larger than twelve members makes it difficult for all to see the same birds, and beginning birders are less likely to receive the attention they require from the group's leader. Whenever possible, organize car pools to minimize congestion and save fuel.

2. *Scouting Trip.* Visit the intended locale the day before the field trip to see what birds you are likely to encounter with your group. A scouting trip may help to avoid embarrassing delays from such logistical snags as closed roads, dried-up marshes, or confusion arising from secondhand travel directions.

3. *Avoid Off Hours.* Schedule birding trips to start as early in the morning as possible. Most land birds are active in the early morning, rest during the

heat of the day, and resume feeding in the late afternoon. This pattern is most evident in spring and early summer when bird song and territorial defense behaviors are conspicuous. (After approximately 11 A.M. birds become less active and hence harder to locate; activity generally increases after 3 P.M.) Night hikes in search of owls, woodcocks, and snipes are unique experiences that are likely to leave positive memories because of the unusual hour—regardless of the success of the search.

4. *Comfort.* Suggest appropriate field clothes and footgear so that children and beginners will know how to dress. Even if birds cooperate, wet feet or chills can easily spoil the trip. For day-long trips be certain to include an occasional rest stop.

5. *Orient the Group.* At the beginning of the trip review these points: (a) Ask if anyone needs assistance with binoculars. If necessary, take a moment to review focusing procedures. If there are members of the group without binoculars, pair them with someone who is willing to share. (b) Obtain permission to enter private property well ahead of time and diplomatically remind group members to respect the rights of property owners. Caution the group to avoid indiscriminate use of binoculars near windows. (c) Birding in a group has the benefit of providing many eyes and many ears to detect the movements and sounds of birds. Each person should alert the leader and other group members when he or she sees a bird. *All* birds are worth reporting. There is no such thing as a "trash" bird. (d) Ask participants to *calmly* describe the location of birds to the leader and other group members using the clock technique or landmark system. Ask participants to help each other locate birds. Insist that descriptions should not contain vague directions, such as "It's over there!" or "It's in that tree!" (e) Review in general terms what birds are expected to be doing at this particular season. Are they setting up territories? feeding young? getting ready to migrate? Alert group members about some of the typical birds they are likely to see (but make no promises!). (f) Explain that the leader should stay at the front of the group. Participants wandering ahead may flush birds that others will miss. If you pick up an item of interest, such as a feather or food plant, move to the center of the group and hold it high for everyone to see.

6. *Set an Example.* Demonstrate good field etiquette by avoiding excessive disturbances to birds and their habitats. Be especially cautious near bird nests, and take extra precautions to avoid interrupting incubation or care of young.

7. *Rarities.* Avoid turning your bird walk into a search for rarities. If there are beginners in the group, they'll be just as excited about common birds as the rare ones. All birds deserve attention; the challenge to the experienced birder is to maintain enthusiasm about the thousandth barn swallow or song sparrow. Remember that there are likely to be people in your group who are seeing the bird for the first time. Try to remember *your* first encounter.

8. *Make Certain That Everyone Gets a View.* Try to let everyone in the group see

all birds. Certain participants will always be the last to see a bird that the rest of the group has easily located. Check to see that such individuals are spotting with their eyes, then lifting their binoculars. Although beginners are often the last to locate birds, their enthusiasm when they finally do spot one is usually ample reward for the leader's extra effort. If you have a tripod-mounted spotting scope, focus it on a bird and encourage everyone to take a look, even if they have seen the same bird with binoculars. At a reasonably close range, scopes deliver impressive views and may leave the participants with long-lasting, vivid images.

9. *Keep Watching.* Keep your binoculars up after you identify birds. What are the birds doing? What behaviors can you identify? What interactions between birds can you see? Have the group sit at the edge of a pond or other interesting location and practice behavior-watching. Good views are more likely to leave enduring memories than are long lists of species.

10. *Report Field Marks That You Really See.* When describing field marks, leaders frequently give the diagnostic marks for a species without actually seeing the marks at the time of a sighting. Unless this is clarified, beginners frequently question their own sighting abilities and assume the leader has extraordinary vision or other extrasensory perception.

11. *Provide Information.* Research the life histories of some common species that you are likely to see. If you encounter them while birding, give interesting details about behavior, life history, and ecology. A brief investment of time reading about common birds will vitalize the trip for everyone concerned. During quiet moments, don't hesitate to inject "nonbird" natural history topics, such as comments about geology, botany, and local human history.

12. *Vary Your Pace.* When there are birds to watch, move slowly, permitting everyone to see. When birds are scarce, quicken the pace to cover ground.

13. *Building to a Climax.* Regardless of how poorly a trip seems to be going, avoid such comments as "You should have been here last week," and avoid keeping the group in the field until their interest is exhausted. Whenever possible, reserve something special for the end of the trip—a glimpse of an active nest, an unusual bird staked out from the scouting trip, or a fine vista of the countryside. If the group comments that it's too bad the trip is over, you'll know that you've ended it at the right time.

14. *Review.* At the end of the trip, pull the group together for a brief review of the birds observed. This is an appropriate time to assemble a group checklist with number estimates for all species. Suggest readings about some of the birds observed during the trip. If the outing was part of an organized field program, encourage participants to join the next trip and bring a friend.

15. *Keep It Fun.* The sport and social elements of birding are basic to its popularity, but the catalyst that really makes it fun is an enthusiastic and considerate field leader.

2

BINOCULARS
AND
SPOTTING SCOPES

SELECTING A BINOCULAR

Although a binocular is the most important tool for watching birds, the selection of the "right" set can be a bewildering task when you consider such features as power, coating, field of view, size, and weight. The selection task is especially difficult because binoculars that look similar from the outside may vary in price from $20 to $600. Here are some tips to keep in mind when making your selection.

Power

Examine the flat upper surface of a binocular housing and you'll note two numbers, such as 6x30, 7x35, or 8x40. The first of the two numbers designates the binocular's power. Seven-power binoculars make subjects appear seven times larger than they would without magnification. Some birders prefer binoculars as powerful as 10x for viewing birds that are likely to be in open situations or birds that are not very active, such as waterfowl and shorebirds. However, most bird-watchers prefer 7x or 8x binoculars because higher magnification is quickly offset

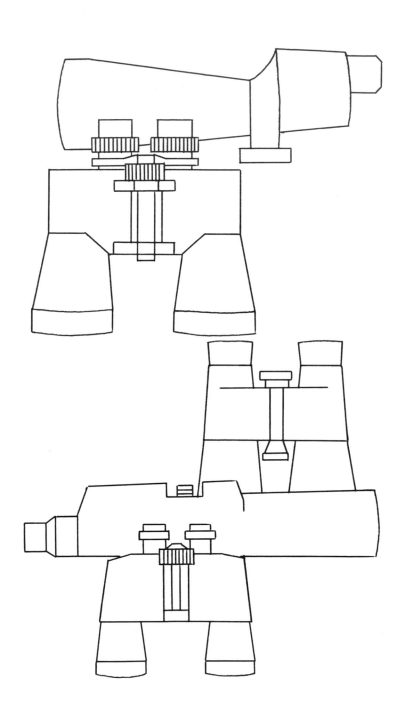

by serious disadvantages. The higher the magnification, the more difficult it is to hold binoculars steady. High-power binoculars not only magnify the subject, but also movements of the observer. The result may be blurred images, eyestrain, and headaches. Further drawbacks of high-power binoculars are darker images, a smaller field of view, and the inability to focus on subjects close to the observer. The zoom feature of certain models offers the capability of quickly increasing from 7x to 15x, but at the higher magnifications such binoculars are too difficult to hold, and the dark images make it nearly impossible to see important field marks.

Light-gathering Capacity

It's as important to have adequate light-gathering capacity as it is to have binoculars that give you a sharp image. Only a bright image will reveal detailed field marks and exhibit the full beauty of bird colors.

Light enters the binocular through the objective lenses. The diameter of these lenses is the second of the numbers on the binoculars, as in the 7x35 designation. Accordingly, 7x50 binoculars have larger objective lenses, but there is no gain in magnification. The advantage of the larger objective lens is its greater light-gathering capacity. Just as the large eyes of an owl gather enough light to permit nocturnal vision, so binoculars with large objective lenses are an advantage for bird-watching at dusk and dawn or in dark habitats such as forests.

The exit pupil is the best measure of a binocular's brightness. You can easily see the exit pupil by holding the binoculars at arm's distance and looking into the

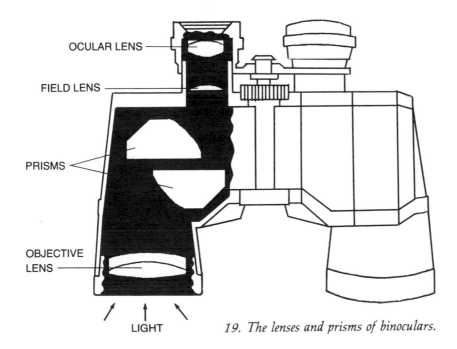

OCULAR LENS

FIELD LENS

PRISMS

OBJECTIVE LENS

LIGHT

19. The lenses and prisms of binoculars.

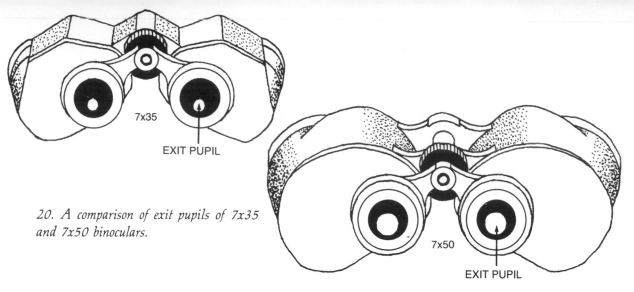

20. *A comparison of exit pupils of 7x35 and 7x50 binoculars.*

eyepieces. Depending on your binoculars, the exit pupil will vary in appearance from a dark pinhole to a brilliant, clear circle. To find the exit pupil size, divide the size of the objective lens by the magnification number. Thus, the 7x35 binocular has an exit pupil of 5mm, as compared to the brighter 7.1mm exit pupil of the 7x50 binocular.

Bird-watchers looking for a binocular to use on rolling boats will find that an exit pupil of at least 5mm offers a distinct advantage. When motion causes your binocular to move in all directions around your eyes, you may experience image blackouts as the exit pupil moves away from the pupil of your eye. In bright daylight, when your eye has a pupil opening of 2mm, a binocular with a 5mm exit pupil provides a 3mm leeway to adjust to the movement.

While binoculars with larger exit pupils are better for boating and generally offer brighter images and better color, these advantages are not achieved without trade-offs. The principal drawback is the additional weight of the objective lenses and the oversized housing necessary to support the larger and heavier optics.

Check the following table to determine which exit pupil size meets your needs:

Exit Pupil Size	*Appropriate Situations*
2–4mm	Bright-light situations (such as open agricultural lands, exposed mountains, shorelines)
4–5mm	Shaded light situations (such as forests); boating
Over 5mm	Dusk and dawn

Certain features may cancel the light-gathering advantages of large binoculars with outsized objective lenses. Carefully examine the edges of the exit pupil to see if it is a complete, bright circle or if the edges are shaded gray, resulting in a bright central square (see illustration 21). If only the center of the exit pupil has full brightness, then some of the light is being blocked by inferior internal optics, and the advantages of large objective lenses will not be fully realized.

21. The exit pupil on the left is blocked by gray at the edges, indicating some light is being blocked by inferior internal optics. The exit pupil on the right is not blocked.

Light entering the objective lens must pass through as many as eight pieces of optical glass in each barrel. At each glass surface (up to sixteen) some light will be reflected backwards rather than passing through the binocular. The optics of well-made binoculars are coated with an even film of magnesium fluoride that helps deliver more than 90 percent of the light gathered by the objective lenses. Without coating, binoculars may reflect away from your eyes up to 60 percent of the light that entered the objective lenses.

Coated optics are also an aid when looking at backlit subjects. Without coating, light reflects within the binocular, causing annoying glare. Even with coated optics, never look directly at the sun, because intense magnification of sunlight can cause eye damage.

Although most manufacturers will coat the exterior lenses, some inexpensive binoculars may not have all interior optics coated. To check lens coatings, hold the binocular under a fluorescent light and look into the objective lenses. Search for purple-violet or amber reflections on all interior optics. White reflections are evidence of uncoated optics. Even though some manufacturers state that their binoculars have "fully coated optics," it's best to check for yourself.

Field of View

Field of view is the width of the image you see while looking through the binoculars. It is usually measured as the width of area visible at 1,000 yards from the observer—for example, 400 feet at 1,000 yards. Sometimes the field of view is expressed in degrees, usually engraved on the surface of the housing near the eyepiece. If so, you can calculate field of view by multiplying the number of degrees by 52.5 (the number of feet in 1 degree at 1,000 yards). Thus, the field of view for a 7 degree binocular would be 7 × 52.5 = 367.5 feet at 1,000 yards.

The wider the field of view, the easier it is to locate birds with your binoculars. Wide-angle binoculars are of special value to beginning bird-watchers, since the larger field makes it easier to find birds—especially flying birds or those inhab-

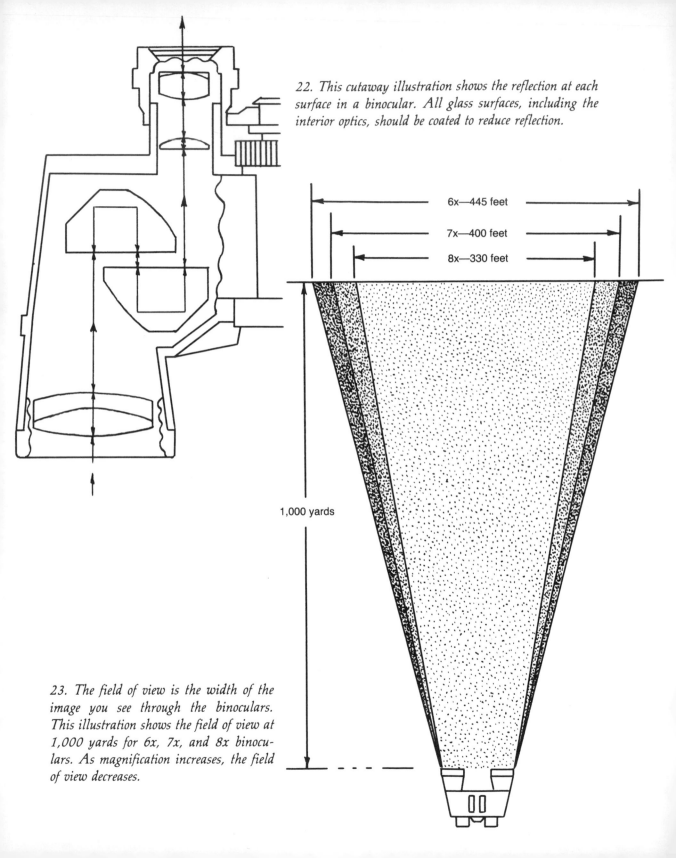

22. This cutaway illustration shows the reflection at each surface in a binocular. All glass surfaces, including the interior optics, should be coated to reduce reflection.

6x—445 feet

7x—400 feet

8x—330 feet

1,000 yards

23. The field of view is the width of the image you see through the binoculars. This illustration shows the field of view at 1,000 yards for 6x, 7x, and 8x binoculars. As magnification increases, the field of view decreases.

iting dense vegetation. Extra-wide-angle binoculars expand the field by increasing the size and number of lenses in the ocular system. The additional optics increase the cost of the binoculars and make them bulky to hold. Since it's difficult to produce binoculars that have sharp images across the entire field, beware of low-cost binoculars with a claim of a wide field of view. They are probably sharp only in the center of the field. With experience, most bird-watchers find that a standard field of view is usually adequate and that there is little need to invest in extra-wide-angle binoculars.

Resolution

High-quality optical glass may cost more than $300 a pound, and each lens and prism must be professionally ground and mounted with expert precision. High-quality binoculars are finely crafted instruments. Attention to such details as balancing and matching lenses is basic to the production of the best products. Some manufacturers cut corners throughout production, especially by using low-quality glass and less of it. For these reasons, the price of a binocular is one of the best indicators of quality.

High-priced binoculars usually have better optics that result in better resolution, as evidenced by sharper images and less eyestrain. Quality binoculars have crisp images from the center to the edge of the field. You can check this by looking at a map or newspaper tacked to a wall. Stand back about 25 feet and see if you can read the print in both the center and edge of the field.

Alignment

The twin-barrel design of binoculars makes them vulnerable to loss of alignment. When binoculars are functioning properly, both sides focus on the same field of view; however, if a binocular receives a sharp jolt, it can easily be thrown out of alignment, and the two fields will no longer overlap. If you look through an out-of-alignment binocular, your eyes will attempt to bring the two views together. If alignment is far off, you'll see double images, and the subject will look blurry. Binoculars that are only slightly out of alignment may be more of a problem, because your eyes will strain to bring the two images together; the result will be fatigued eyes and a headache.

Low-priced binoculars are more likely to develop alignment problems than higher-priced models. Prisms and lenses in cut-rate models may be glued in place rather than securely strapped by metal brackets. Temperature changes or slight jars can throw inexpensive binoculars out of alignment. Realigning binoculars is not a simple task, as they must be taken apart by an experienced technician and recalibrated using special equipment. It may be less expensive to replace a binocular than to have it realigned. Higher-priced binoculars are more likely to withstand the stress of continued field use, and, if treated with reasonable care, they should last a lifetime.

Because even new binoculars may be out of alignment, it's important to occasionally check the alignment. Try the simple test in illustration 24.

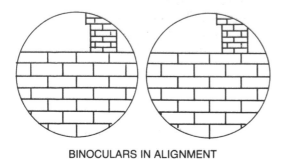

BINOCULARS IN ALIGNMENT

BINOCULARS OUT OF ALIGNMENT

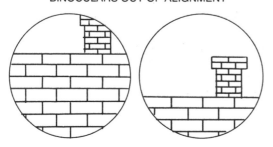

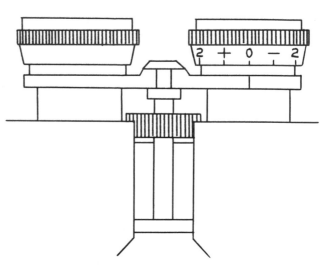

24. Check the alignment of new binoculars by looking at the roof of a house through them, then move the binoculars about 8 inches away from your eyes. If the binoculars are in alignment, the horizontal line of the roof should be at the same level in both fields.

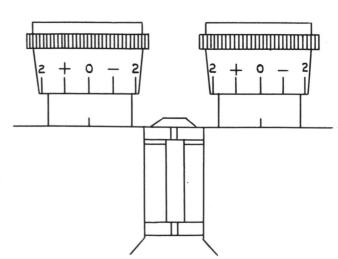

25. Center-focus binoculars (top) and individual-focus binoculars.

Center Focus Versus Individual Focus Binoculars

Most bird-watchers prefer center focus binoculars to individual focus eyepieces because the center focus design permits the viewer to focus on birds that are as close as 10 feet. In contrast, individual focus binoculars must be reset when looking at subjects closer than about 30 feet, which makes them awkward for watching most land birds. The individual focus design is better suited for use on boats, where birds seldom come closer than 30 feet and the sealed nature of the design gives important protection from rain and water spray.

Roof Prism Binoculars

Roof prism binoculars have straight barrels, a feature achieved by placing the two prisms in each barrel close together. Roof prism binoculars offer several advantages over the standard (Porro) prism design. Roof prism models are compact and lightweight, and have excellent resolution without sacrificing brightness or field of view. Internal focusing mechanics have the further advantage of being well sealed from moisture and dirt.

Roof prism binoculars do not offer the depth perception of the standard Porro prism design, nor do they focus as closely. Most roof prism binoculars require frequent fine focus adjustment for changes in depth of field. Their principal drawback, however, is the high cost for repairs and initial purchase price, which may exceed $600.

26. Roof prism binoculars.

Mini Binoculars

Palm-sized binocular midgets are gaining increasing popularity among bird-watchers. There are more than forty different models available, with prices ranging from $19.95 to $350. Mini binoculars are the ultimate in lightweight compactness and the better models deliver images of sharpness comparable with that of standard (porro prism) and roof-prism binoculars. The principle trade-off is light. With objective lens sizes no larger than 30mm, these tiny binoculars are most useful in bright light situations.

Models within the upper price range are finely crafted instruments containing more than 100 individual mechanical parts and 16 optical components. Tested to withstand the rigors of temperature extremes and sudden jolts, they provide a viable binocular option to the bird-watcher already encumbered by bulky camera gear, tape-recording equipment, and field guides.

Binoculars for Eyeglass Wearers

People who wear eyeglasses should focus their binoculars through their glasses so that they can lift the binoculars directly to their eyes without first removing their eyeglasses. This creates a special problem, because eyeglasses block binoculars from close contact with the eye, and the extra distance between eye and lens reduces the field of view. Shallow eyecaps on your binoculars can help solve this problem. Some binoculars are available with interchangeable shallow and deep eyecaps. Rubber eyecaps are another approach to the problem. These roll down for use with eyeglasses and will not scratch eyeglass lenses. Binoculars that are specially designed for eyeglass wearers often have the letter B following the numbers of the objective lens—for example, 8x42B.

How to Shop for Binoculars

After narrowing your choice regarding power, objective size, and field of view, perform the following tests on that bewildering collection of binoculars behind the store counter. Save your final decision regarding price until after you've examined what's available.

1. Compare binoculars of the same magnification by holding one atop the other. Outside or in adequate store light, alternately look through each binocular, comparing them for brightness and clarity. Compare the best from your first comparison with a third—each time choosing the binocular with the best characteristics. Continue this process of elimination until you've examined what's available.
2. Holding the binoculars at arm's distance, check the exit pupils to see if they are blocked by gray shadows at the edges. Nearly all binoculars under $100 will show the gray border obstructing the exit pupil.
3. Look into the objective lenses to see that all lens surfaces are coated with an even purple-violet or amber hue. Carefully examine the objective and ocular lenses for scratches.

4. Be sure that all mechanical parts move smoothly and that the bridge supporting the barrels does not wobble.

5. Outside the store, check alignment by looking at a rooftop or horizontal power line. Carefully examine the print on a sign to see if you can read the lettering at the edge of the field as well as at the center.

6. Look at the edge of a backlit sign or building to see if there is a band of bright color fringing the object. This is the result of an inferior lens system unable to bring light of different wavelengths together at the same point.

After narrowing the field to a few choices, select the highest-price binoculars you can afford (see Appendix A: Sources for Binoculars). Price is your best measure of craftsmanship and materials. Lower-priced binoculars require compromises, but even an inexpensive pair will launch your bird-watching experiences. You can always retire your first pair to the status of a backup or share it with a beginning bird-watcher.

FOCUSING BINOCULARS

To focus binoculars, stand about 30 feet away from a sign with clear lettering and follow these steps.

1. Note that the two binocular barrels pivot on the hinge post, permitting the eyepieces to fit your eyes comfortably. Facing the sign, spread the barrels as wide as you can. Then, with the binoculars held at eye level, press the barrels together until the two images converge into one. The reading on the hinge post will remain constant for your eyes.

2. Turn the center focus wheel as far to the right as you can, thus moving both eyepieces to the maximum distance from the body of the binocular. Note that only one of the eyepieces has calibrations on it (usually the right eyepiece). Turn the calibrated eyepiece in a counterclockwise direction, moving it as far out from the body of the binoculars as possible. Now both eyepieces should be out of focus.

3. Facing the sign with eyes shut, lift the binoculars into position. With your left eye open, turn the center focus wheel until the lettering comes into sharp focus. To be sure you have the sharpest possible focus, pass the sharpest point and then back up to find it again.

4. Shut your left eye and with your right eye open, turn the right eyepiece in a clockwise direction, bringing the lettering into focus. Pass the point of sharp focus and then back up to where the lettering is sharpest. Open both eyes, and the binoculars will be in perfect focus.

5. Take note of the right eyepiece setting, as it is now adjusted to your eyes. The only reason for it to change would be to accommodate a change in vision. Some people tape the right eyepiece in place to keep it from shifting accidentally. After the right eyepiece is set, you need only adjust the center

wheel to focus both eyepieces. For individual focus binoculars, focus both right and left eyepieces as described above for the calibrated eyepiece.

BINOCULAR CARE

Binoculars used for bird-watching face many hazards to which binoculars used at opera houses and football stadiums are rarely subjected. In addition to being able to withstand precipitation and highly corrosive salt spray, bird-watching binoculars should be rugged enough to accompany birders up rocky slopes, in and out of boats, down sandy beaches, and through wet and dry bird habitats. Center focus binoculars (the preferred type for birding) are particularly vulnerable to water and dirt, which may enter through the focusing apparatus. Dirty binoculars provide neither sharp detail nor crisp colors.

How to Clean Binoculars

Binoculars should be cleaned frequently, following these suggestions.

1. Thoroughly wipe dirt off metal parts and brush all lenses with a wad of lens cleaning tissue or a soft camel-hair brush to dislodge particles of sand and grit. Unless these are removed, you could easily scratch the lens and its coating during the cleaning process. Hold binoculars upside down so that dirt will fall away from the lens surface.
2. Fold a piece of lens cleaning tissue so that it is at least four layers thick. This prevents oil from your fingers soaking through the tissue and onto the lens surface. Use a circular movement to gently wipe all lens surfaces.
3. If there is a film of oil on the lens, put a drop of lens cleaner on the tissue and repeat the circular wiping movement.
4. Look for dirt on internal optics by holding the binoculars up to a light and looking into the objective lenses. Never attempt to open binoculars, since alignment is easily disrupted. Though it's expensive, leave internal cleaning to the professionals.

Protecting Binoculars

1. Don't stroll through the woods swinging binoculars by the strap. When in the field, always keep the binoculars around your neck.
2. When performing active maneuvers, such as jumping across a ditch, climbing a rock slope, or getting into a boat, tuck the binoculars inside your jacket or secure them under your arm.
3. When driving in a car, don't leave the binoculars on the seat, since a quick stop will send them flying forward. Be certain not to leave your binoculars sitting exposed in your car, especially on a hot summer day. If thieves don't find them, the sun may soften the lens coatings, causing them to crack and separate from the lenses.

4. Keep binoculars under cover if it starts raining. Water can leak into the housing, carrying with it dirt that will stain the internal optics. Rain guards offer some protection during light rains and drizzles, but they are not adequate protection for heavy rains. If your binoculars "steam up" on the inside, set them in a warm, dry place, and they are likely to dry out in a couple of days. Otherwise, fungus may start growing on the lens coating.

5. If binoculars fall into fresh water, have them professionally cleaned as soon as possible to avoid rusting. If they are dropped in salt water, rinse them thoroughly in fresh water, seal them in a plastic bag, and rush them to a professional service department within three days of the soaking. If the water-soaked binoculars were inexpensive to start with, you might as well buy a new pair, because the repair charge will probably exceed the purchase price.

SELECTING A SPOTTING SCOPE AND TRIPOD

Spotting scopes provide the magnification necessary to see distant birds and to admire the detail at closer ranges. The problems of less light and more vibrations that accompany power increases with binoculars also apply to spotting scopes. High powers magnify the air as well as the subject, often producing hazy images or distracting shimmering from heat vibrations over water and flat expanses such as farmlands. Under good observation conditions, the large objective lens and the fact that scopes are used with tripods permit telescopes to greatly magnify images beyond what you would see with binoculars.

Zoom lenses offer the convenience of changing powers from 20x to 45x by simply making a single adjustment. Other spotting scopes have interchangeable eyepieces up to 60x or more, but viewing conditions are seldom good enough to use more than a 45x eyepiece. Twenty power is the best all-purpose magnification.

Spotting scopes with 15x or 20x eyepieces can be mounted on a modified rifle stock to eliminate the bulk of a cumbersome tripod. A home-built stock can be carved from a block of basswood or sugar pine and molded to individual comfort. To custom build a spotting scope stock, first make a cardboard pattern and trim it so the eyepiece comfortably fits your eye. Commercially built stocks for spotting scopes and telephoto lenses are also available.

A rifle-stock-mounted scope is difficult to share with a group of bird-watchers. In contrast, a scope mounted on a tripod is more likely to give everyone in the group a good view. Some tripods are clumsy to use because they may have as many as nine different locks and clamps controlling the extension of the legs. Not infrequently, the last leg is secured just as the bird leaves. The most efficient tripods are those that feature "flip locks." These tripods are easy to operate because once the legs are released, they fall to their own level and are locked in place by the three locks conveniently located at the top of each leg.

Scopes are most frequently used to watch slow-moving birds, such as waterfowl, shorebirds, and hawks, that live in expansive habitats such as wetlands and open fields. Scopes can also permit full-field, intimate views of birds already at

27. Spotting scope mounted on a custom-built stock.

28. Birder using a scope mounted on a flip-lock tripod.

close range. Such scope-aided views frequently reveal the intricate beauty of a bird's plumage and permit the observation of behavior that might otherwise go unseen.

Birding with a Small Reflector Telescope

Small telescopes are deluxe aids for watching birds. They can also double as acceptable telephoto lenses in the 1,000 to 2,000mm range and are ideal for astronomical use. The principal advantage of the reflector telescope is that it offers significantly greater magnification than spotting scopes without serious loss of light and sharpness.

The Field Model Questar is the best of the small telescopes for bird-watching. This telescope gathers light through an 89mm objective lens (compared to the 60mm diameter of most spotting scopes). The telescope weighs about as much as a spotting scope (approximately 3½ pounds) and is only 8 inches long. Its ability to focus from distant horizons to 10 feet away gives it excellent flexibility as a bird-watching tool.

Scanning is accomplished through the built-in viewfinder, which provides magnifications of 4x or 8x, depending on the eyepiece used. Small flip levers engage either the medium- or high-power magnifications. Several eyepieces are available, but the 32mm eyepiece is the most useful focal length, since it permits magnifications of 40x and 64x that are a marvel of brilliance and sharpness even under the darker conditions of dawn and dusk. Magnifications as great as 260x

29. Birding with a Questar.

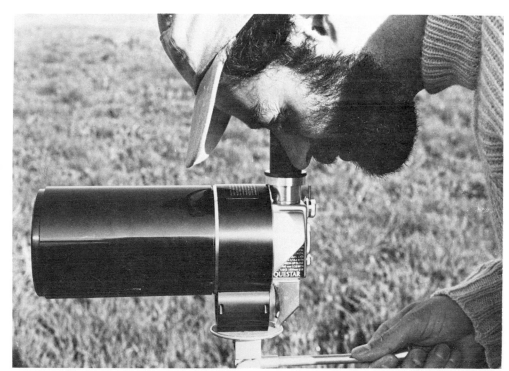

are achieved with interchangeable eyepieces, but such high magnifications are accompanied by some loss of light and a reduction of field of view.

The greatest drawback to the Questar and other small telescopes of the mixed lens/mirror design is the high purchase price, which may exceed $1,500. For birders who can afford them, they offer an unequaled extension of vision for watching birds.

How to Shop for a Bird-watching Telescope

When selecting a telescope, consider how you will use the scope. Do you want a scope primarily for viewing birds, or one that will be adaptable for photographic and astronomical use as well? Spotting scopes are rugged field instruments of moderate cost that are best used just for watching birds. If you want to use the telescope frequently for photography and/or astronomy, then the more expensive (and delicate) reflector telescopes may be the answer. Here are some suggestions to consider when making your selection.

1. 20x is the best all-purpose eyepiece for spotting scopes. Due to distortion and loss of light, eyepieces larger than 45x are usually useless.
2. Zoom lenses that vary in power from 20x to 45x are ideal for most bird-watching. They permit convenient scanning at low power and then a quick shift to higher power for looking at details. If you plan to use more than one power objective, purchase either a zoom or a turret swivel head, as changing screw-mounted eyepieces in the field is awkward and time consuming.
3. Don't buy a cheap telescope. Inexpensive scopes deliver fuzzy, distorted images. Manufacturing shortcuts will only lead to the early demise of your instrument and disappointing field performances.
4. Select a rigid tripod with as few leg adjustments as possible. The flip-lock design provides a secure mount for your scope and a quick way to set the legs on uneven terrain.

Telescopes offer to expand your vision at least three times beyond that of binoculars. You'll be amazed what a difference that makes.

INSURING FIELD EQUIPMENT

Considering the perils to which bird-watching binoculars and other field equipment are exposed, it's a good idea to insure them against damage or loss. If you have homeowner's or renter's insurance, ask your agent to extend the policy to give all-risk worldwide coverage to your binoculars, scopes, cameras, and other equipment. The basic renter's policy will insure equipment at home and in the field against presumed theft, but for a few dollars extra, the all-risk coverage also protects your equipment if it is damaged, lost, dropped overboard, or run over by a passing tractor.

CHAPTER

3

―――――――――――

OBSERVING
BIRDS

WATCHING BIRD BEHAVIOR

Behavior watching adds a new dimension to birding, because it shifts the emphasis from identification of species to observation of individual birds. Even the most common birds can be entertaining as they reveal their repertoire of behaviors. The more familiar you become with a species, the more complete will be your appreciation of how neatly every behavior answers some demand of the bird's habitat. Still further satisfaction awaits observers who know a particular species well enough to distinguish personalities of individual birds.

Keep watching after you've established identities and you'll find that birds are usually busy with a variety of interesting behaviors. New bird species are seldom discovered, but even the most abundant birds perform activities that are either unknown or seldom described. Even the most common activities, such as feeding behaviors, need further attention. For example, certain species are known to consume some of their wing and tail feathers during molt, but this activity is seldom reported from the field. Similarly, although it is widely recognized that carnivorous hawks and owls regurgitate pellets of nondigestible materials such as fur and bones, there is also evidence that some insect-eating birds such as warblers also cough up pellets of nondigestible food parts—yet little is known about the frequency of the behavior or how widespread it is among insectivorous birds.

When watching bird behavior, remember that bird senses are quite different from our own. While most birds have remarkably sharp eyes but a poor or nonexistent sense of smell, certain species have sensory talents that are nothing short of astounding by human standards. Recent studies have shown that birds can orient themselves by star patterns and magnetic fields and that they are capable of hearing extremely low-frequency sounds known as infrasounds. The capability to detect infrasounds may give birds a remarkable capacity to hear sounds from far-distant sources. This ability suggests that a bird living in the central United States could conceivably hear the sounds of both the Atlantic and the Pacific!

Since birds often respond to situations at a highly instinctive level and their senses are so different from our own, it is essential that behavior watchers avoid assigning human motives and values to bird behavior. Birds are neither good nor bad, and, despite perceived similarities in their behavior to that of humans, they are not sneaky, happy, or sad. Such descriptions are difficult enough to understand when applied to humans, but they are totally inappropriate for describing bird behavior. The next time you go birding, focus your attention on bird behavior and see how many of the following behaviors you can observe.

Body Care Behaviors

Because all birds have feathers, it is not surprising to find widespread similarities in the way they maintain their plumage and care for their bodies. Birds ranging in size from eagles to hummingbirds will bathe and preen in surprisingly similar ways, yet careful observation will often show subtle differences between even closely related species. Start watching for the different body care behaviors of backyard and feeder birds; most behaviors in this category are easily observed at any time of the year.

30. **Preening.** Birds must devote a large portion of each day to arranging their plumage. Preening typically consists of sliding each feather through the beak, nibbling the feather from base to tip so that separated feather vanes are zipped together.

31. Oiling. *Most birds keep their feathers lubricated and waterproofed by spreading oil from their preen (uropygial) gland. The gland, located at the base of the tail, is usually stimulated by a squeeze from the bird's beak. The oil is spread onto the plumage and worked into the feathers by preening.*

32. Bathing. *Bathing helps to reorganize the plumage more than to clean it and is usually followed by active preening and spreading of oil from the preen gland. Most land and water birds bathe. Watch carefully to see which parts of the body are wet first, and see if you can discover a predictable pattern.*

33. Dusting. *Birds of dry habitats perform bathing behaviors in depressions scooped out of dusty soil. Dusting may help birds control ectoparasites such as feather mites. Some birds, such as hawks, kinglets, and sparrows, bathe in both dust and water.*

34. Anting. *Many land birds sit on active ant colonies, permitting the ants to roam freely through their plumage. Some species lift their feathers, permitting the ants to crawl to their skin, while others pick up the ants and, with a squeeze, tuck them into their plumage. The behavior is little understood, but it may help to reduce ectoparasites by treating the feathers with formic acid from the ants.*

35. Bill wiping. *After eating, land birds usually clean excess food from their beaks by wiping one side of the beak and then the other against a convenient perch. Bill wiping is commonly expressed at moments of tension, even if the bird has not eaten recently.*

36. Panting. *Birds pant to cool themselves when frightened or overheated. Panting is usually accompanied by gular fluttering, a rapid quivering of the throat that promotes cooling.*

37. Feather ruffling and fluffing. *Feather ruffling is a cooling behavior that permits warm air near the skin surface to escape when feathers are raised. Although similar in appearance to ruffling, feather fluffing occurs during cold temperatures and conserves body heat by increasing the insulating function of the plumage.*

38. Feather settling. *When arranging their plumage, birds raise their feathers, shake their bodies, and stretch their wings before settling the feathers back into place. Unlike feather ruffling or feather fluffing postures, which are prolonged responses to temperature stresses, feather settling is not temperature related and lasts only a few seconds.*

39. Stretching. *Birds frequently stretch the limbs of one side of their bodies. Typically, the wing, leg, and half the tail are extended out and backward. After the stretch is completed, a bird will frequently repeat the process with the other side of its body.*

40. Sleeping. *Most birds have a headless appearance when sleeping because they lay their heads on their backs, tucking their beaks under the shoulder feathers. Birds that are resting, but not sleeping, may shut their eyes, but they do not usually put their beaks on their backs. Birds often yawn before resting or sleeping.*

41. **Scratching.** *Birds scratch with their feet in predictable ways. Most perching birds scratch over their wings, but some closely related species scratch in surprisingly different, yet also predictable, ways.*

Feeding Strategies

Although special adaptations set limits on the kinds of foods birds can eat, most birds consume a surprising variety of food, frequently switching feeding strategies depending on what foods are available. Even birds that have beaks especially adapted for specific foods can often switch to alternates. Hummingbirds, for example, frequently prey upon the small insects they find in flowers, although their primary food is flower nectar. Grosbeaks, which normally use their massive beaks to crack seeds, primarily feed their young a diet of easy-to-digest, high-protein insects and earthworms. Any dramatic changes in diet usually require impressive shifts in feeding strategies.

Carefully observe opportunistic foragers such as gulls, crows, and starlings, which use their all-purpose beak and body form to consume a remarkable variety of plant and animal foods. Also watch for subtle differences in the places birds forage. Studies of warbler feeding behavior show that several species can forage on the same tree without competing for food by feeding at various distances from the trunk to the outermost branches. Likewise, different shorebird species feed by foraging at varying distances from the shoreline, and woodpeckers find different prey in the same tree by chiseling holes of various depths. On your next birding trip, see how many feeding strategies you can observe.

42. Hawking. *Searching for food while perched, then flying out to capture flying insects, and returning to the same or a nearby perch. Examples: kingbirds, woodpeckers, waxwings.*

43. Perch gleaning. *Searching for prey while perched in a tree or shrub and capturing insects or other small invertebrates, such as spiders, mites, or ticks, without flying from the searching position. Examples: most wood warblers, chickadees, titmice.*

44. Sally gleaning. *Searching for food while perched, then flying out to snatch an insect or other small invertebrate prey off a distant surface, such as leaves or a tree trunk. Examples: red-eyed vireo, least, Acadian, and willow flycatchers.*

45. Hover gleaning. *Hovering while searching for insects or other small invertebrates on tree leaves, a tree trunk, or other surfaces.* **Examples:** *kinglets, great-crested flycatchers, phoebes.*

46. Chiseling. *Pounding on a tree trunk creates a cavity and exposes the prey. Pounding also disturbs insects and other small invertebrates, making them come to the surface.* **Example:** *woodpeckers.*

47. Probing. *Reaching into tree bark, soil, or mud with the beak in search of prey.* **Examples:** *brown creepers, nuthatches, woodcocks, dowitchers.*

48. Leaf tossing. *Scratching at the leaf litter layer with the feet, looking for prey under leaves.* **Examples:** *towhees, fox sparrows, ruffed grouse.*

49. **Sweeping.** *Searching for insects in the air and capturing them in flight with a wide-open mouth.* **Examples:** *swallows, swifts, nighthawks.*

50. **Stooping.** *Dropping at great speed in pursuit of flying birds or insect prey.* Example: *falcons.*

51. **Pouncing.** *Flying to the ground to capture prey.* Examples: *owls, hawks, eagles.*

52. Shell smashing. *Dropping clams, mussels, or turtles from the air onto land surfaces to crack the shells and obtain the contents.* Examples: *gulls, crows, eagles.*

53. Surface feeding. *Capturing prey by skimming or plucking food from the surface of the water.* Examples: *black skimmers, petrels, frigate birds, gulls.*

54. Plunging. *Diving from the air or a perch and completely submerging under water in pursuit of fish.* Examples: *ospreys, kingfishers, common terns, gannets, brown pelicans.*

55. Diving. *Submerging from the water surface in pursuit of fish. Examples: loons, diving ducks, grebes, auks, cormorants, and anhingas.*

56. Stalking. *Standing, walking, or wading in search of fish, insects, or other prey, or snatching animals from shallow water or the land surface. Examples: herons, gulls, plovers, robins, thrushes, larks, dippers.*

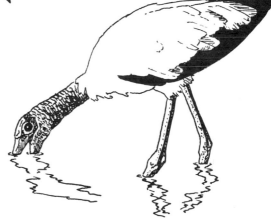

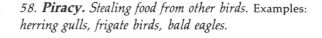

57. Foot raking and stamping. *Disturbing hidden prey in bottom sediments of shallow water by raking mud with the foot or stamping on moist soil to disturb earthworms. Examples: herons, egrets, storks, cranes.*

58. Piracy. *Stealing food from other birds. Examples: herring gulls, frigate birds, bald eagles.*

59. Scavenging. *Consuming dead animal matter, usually searching from the air and then feeding on the ground or water.* Examples: *vultures, crows, gulls.*

60. Sifting. *Straining animals and plants from mud or water.* Examples: *flamingos, spoonbills.*

61. Grazing. *Biting off or pulling up rooted land vegetation.* Example: *geese.*

62. Dabbling. *Tipping tail up and stretching downward without completely submerging to reach the bottom vegetation of shallow waters.* Examples: *swans, mallards, and other dabbling ducks.*

63. Pecking. *Searching on the ground and picking up loose seeds or consuming fallen fruit.* Examples: *geese, pheasants, juncos, sparrows.*

64. Pruning. *Nipping off and consuming twigs and buds.* Examples: *grouse, grosbeaks.*

65. Plucking. *Removing fruit or seeds from vegetation.* Examples: *robins, cedar waxwings, purple finches, goldfinches.*

66. Nectar hovering. *Hovering at flowers to feed on nectar.* Example: *hummingbirds.*

Social Displays

Social displays include all interactions between birds of the same species. Certain displays, such as mobbing and flocking, appear in a great variety of birds, while other displays, such as song and courtship performances, are unique for each species.

Most social displays are linked in some way to the reproductive cycle and are greatly influenced by seasons of the year. When watching social displays, look for ritualized body care behaviors (such as preening and feather-ruffling postures). Such behaviors are often incorporated into courtship performances, but they can surface at seemingly inappropriate times, when the bird is under stress, and are then known as "displacement activities" (such as bill wiping).

*67. **Threat and appeasement displays.** Birds frequently exhibit threat displays when defending feeding or nesting territories from intruders. Common elements of the threat display (left) include gaping (opening beak as if to bite) and wing opening and head lowering (as if to attack). Actual fights among birds are rare because of the effectiveness of such displays. Birds usually assume an appeasement posture when retreating from a threat. A bird in an appeasement posture (right) frequently hunches its shoulders, lowers its beak, and turns away from the threatening bird. Watch for threat and appeasement displays at feeding stations.*

*68. **Caution displays.** When approaching their nest, birds often assume a nervous caution display consisting of searching in an erect posture, tail flicking, and bill wiping. The duration of caution displays is greatly influenced by the presence of predators and human observers.*

69. Mobbing. *When a predator, such as a fox, snake, or owl, is discovered, birds of many species will join together to chase the predator from the vicinity.*

70. Flocking. *Many species flock together prior to and during migration. Other species roost together at night, gaining protection from predators and perhaps communicating information about available food supplies. Even in dense roosting flocks, birds always maintain an individual distance of at least several inches from the nearest bird.*

71. Calls. *Given by both sexes, calls are short, nonmusical vocalizations that function to communicate information about daily activities, such as feeding, flocking, migration, and alarm.*

72. Songs. *Each species has a unique song that the male uses to attract a mate and to alert other males of the same species that the territory is occupied. With few exceptions, female birds do not sing.*

73. Courtship displays. *Unique combinations of body postures and distinctive plumages function to attract a mate of the same species. Courtship displays stimulate the female to lay eggs and help to keep the pair synchronized and together until the breeding cycle is complete.*

74. Courtship feeding. *In this courtship display, the male feeds the female, who typically responds by quivering her wings and accepting the food much like a begging nestling.*

75. Mounting and copulation. *A male usually mounts the female for a few seconds by standing on her back, balancing himself with wings partly spread and body fluffed. The female often takes a submissive posture with sleeked plumage and quivers her wings. Because actual copulation is often difficult to confirm—as the male tucks his tail to the side and under the female's tail—it is best for birders to distinguish between mounting and copulation.*

76. Nest building. *A bird gathering or flying with nonedible materials, such as mud, sticks, or feathers, is most likely in the process of building a nest.*

77. Incubation. *Most birds incubate their eggs against a patch of skin, known as the brood patch, especially bared by molting during the breeding season. Incubation takes about two weeks for small land birds.*

78. Egg turning. *Parent birds must turn the eggs several times each day if the embryos are to develop successfully. The eggs are usually turned by placing the bill on the far side of the egg and lifting, thus turning the egg in place.*

79. Brooding. *Newly hatched young are not warm-blooded like their parents and must be brooded for several days until they can maintain their own body temperature. At night and during cool days, the parent will warm the nestlings by sitting in the incubation posture. On hot days, parents may stand over the young to provide shade.*

80. Nest cleaning. *To keep the nest clean of parasites and safe from predators, many birds remove eggshells after hatching. The droppings of the young of most small land birds emerge in "portable" casings called fecal sacs. Parents usually remove the sacs as they emerge from the cloaca of the nestlings and either consume them or carry them away from the nest.*

81. Carrying food. *When you see birds carrying food, it is usually evidence that they have a nearby nest with young.*

82. Freezing. *Ground-nesting birds, such as the woodcock, grouse, and certain sparrows, rely on their camouflage and the behavior of freezing when a predator threatens. If a bird flushes from underfoot, be careful, since its nest is likely to be very near.*

*83. **Distraction displays.** When the nest or young are threatened, some parent birds experience a conflict between attacking and fleeing. Birds nesting on the ground in open habitats frequently respond with a compromise behavior called the broken wing act, in which the parent flees the nest but flutters and calls as if it is mortally crippled. The predator usually follows the distressed parent only to find the adult "recovers" at a safe distance from the nest. Birds nesting on the ground in dense, grassy habitats may perform a similar distraction display called rodent running, in which they creep along the ground away from the nest like a vulnerable rodent.*

BEHAVIOR-WATCHING ACTIVITIES

Building an Ethogram

Derived from ethology, the study of animal behavior, an ethogram is a precise catalog of all the behavior patterns of an animal. By carefully watching resident birds throughout the year, you can collect a list of behaviors for each species and assemble ethograms for local breeding birds. The behavioral repertoire of non-breeding birds is not as great, and they are more difficult to keep under observation. However, whether watching birds during the breeding season, migration, or winter, there is always plenty to see.

Start by selecting one resident bird and record your field observations, following the note-taking format described for "Species Accounts" on page 65. While watching for the behaviors described on the preceding pages, see how many behaviors you can observe for your species.

Consult the literature before starting your field observations. (See Chapter 8, "Building a Bird-Watcher's Library," for a complete, annotated bibliography.) *Life Histories of North American Birds* by Arthur Cleveland Bent is the best place to start; you will find pages of interesting and colorful accounts about nearly all North American birds. *A Guide to the Behavior of Common Birds* by Donald W. Stokes is another excellent reference detailing the social displays of twenty-five North American species. However, the behavior of many birds is so poorly known that even after a search of ornithological journals, you may find few studies about the species you are observing.

After reviewing the literature, take field notes, listing and describing the behaviors you see. In addition to the stereotyped behaviors common to many species, such as preening, bathing, and dusting, watch for behaviors and social displays that may be unique to your species. If a name doesn't exist in the literature, give the behavior a short descriptive term based on the most conspicuous action within its limits. Then describe the display in detail. In the following descriptions of the behavior of the great blue heron,[1] note that the displays consist of a predictable sequence of actions (in *italics*) and that the name of the behavior is derived from the most conspicuous action in the display.

> Twig Shake: The heron *extends* its neck slowly, *grasps* a branch in its mandibles, and *shakes* it side-to-side or forwards-and-backwards.
>
> Arched Neck: Quickly the heron *erects* its plumes (crest, scapular, and basal portions of the neck) and *curves* its neck like a rainbow so that the closed bill usually *points* below horizontal. It *maintains* this position up to 5 sec. before *relaxing* to a standing posture.

Other great blue heron courtship display names include Circle Flight, Crest Raising, Fluffed Neck, Bill Duel, Bill Clappering, and Stretch. Names for great blue heron vocalizations, such as Landing Call and Aerial Call, indicate the predictable circumstance during which the vocalization is produced.

Avoid adding interpretations about function to display names. Such adjectives may bias other observers from making independent observations. Names such as Aggressive Head Lowering or Distress Call suggest that the meaning of the behavior or vocalization is clearly known when, in fact, the circumstance and value of the display may be unknown or in dispute. If you have an idea about function, survival value, or origin of a display, include this in a brief statement under your description.

> Twig Shake: The heron *extends* . . .
>
> > Discussion: The variability found in the Twig Shake's form may enable it to carry a large amount of "graded" information: Vigorous performances may well express redirected aggression. The Twig Shake presumably evolved from nest-building motor patterns, which it still resembles.

Build an ethogram by assembling a list of behaviors from your field notes. Use preferred names from the bird literature or names of your own invention, listing the behaviors you see under the two major categories: body care behaviors (include feeding strategies) and social displays. For each behavior include a detailed

[1] D. W. Mock, "Pair-Formation Displays in the Great Blue Heron," *Wilson Bulletin,* vol. 88, no. 2, pp. 185–230.

description of the actions within the behavior or display followed by a separate discussion paragraph if you have an interpretation about the display.

Identifying a Behavioral Sequence

Bird behavior usually follows a fairly predictable sequence of events depending on the season, time of year, and weather. This sequence becomes readily apparent if you concentrate on the activities of an individual bird. Select a bird that is attending a nest or a large waterbird that is likely to stay within your field of view for most of a three-hour period. Record the behaviors you see in field note format, then assign a behavior category to each observation.

In a study of the house sparrow,[2] the following categories were assigned to the activities of a male house sparrow that was busy attending its young:

A   Agonistic Behaviors (threat and appeasement displays)
C   Caution Behaviors (alert postures)
F   Foraging (collecting food for self and nestlings)
N   Feeding Nestlings (entering and delivering food at the nest)
B   Body Care Behaviors (preening, bathing, dusting)

The behavioral sequence emerges by assigning a category to each observation in the field notes (see illustration 84). The following is a sample of the sequence of behaviors performed by the male house sparrow:

A (F C N) A (F C N) (F C N) B (F C N) (F C N) (F C N) B (F C N) B

The routine of this sparrow is apparent from the above sequence. This bird was clearly committed to the routine of foraging for food, cautiously approaching the nest, then feeding the young (FCN). This was interrupted only by threats to the nest from potential predators (A) and occasional preening and bathing (B).

Clocking a Time Budget

In addition to demonstrating predictable sequences in bird behavior, your field notes can also show how much time birds spend on each activity. To construct a time budget, simply total the time for all observations within each behavioral category. During three hours of observation, the male house sparrow used his time as follows:

Behavior Category	Time (minutes)	% of Total Time
A Agonistic	32	18
C Caution	40	22
F Foraging	68	38
N Feeding Nestlings	9	5
B Body Care	31	17
	180	100

[2] D. DeLuca, "Behavior of the House Sparrow (*Passer domesticus*)" (1979). Unpublished Cornell University report.

D. DeLuca
'79

BEHAVIOR	TIME	

HOUSE SPARROW
(*Passer domesticus*)

17 July

Continued

| A | 10 MIN. | **0604** - Male house sparrow situated in sugar maple to west and directly opposite nest. Maples are open enough so that I can see him - he chirps continuously to help with location. Blue jay alights in maple within 20 ft. of nest and sets up a raucous jay-jay-jay alarm call. Cannot determine source of jay's distress. With jay present, the house sparrow takes up station in another maple 10 ft. from nest. He assumes a stretched out position, feathers sleeked down, head and neck stretched out and bill pointed slightly above the horizontal. Tail flicks continuously and chirps in low, rusty manner - in pairs and triplets. Moves to another branch and back. Always near nest. |

| F | 3 MIN. | **0614** - Leaves branch. Flies to open lawn. Hops along through grass, his head bobs constantly, pauses frequently, cocks head from side to side and pecks at grass. Hopping along southern border of rose, lilacs and berry bushes. Head cocked and directed upwards, makes a swift lunge upwards to the leaves, then returns to ground. |

| C | 2 MIN. | **0617** - Returns to nest area. Perches on drain pipe next to nest. Erect position, feathers smoothed down, tail flicking constantly, and head moving in a searching manner. Three, bright green catepillars hang from between his mandibles. Erratic, soft chirping begins. |

| N | 30 SEC. | **0619** - Darts out from pipe and returns. Then darts quickly into the nest between rafters. Female is now located on a branch near nest. Young can be heard to increase the intensity of cheeps as male enters. Male emerges from nest without food 30 sec. laters |

84. *A sample page of field notes on the house sparrow, showing behavioral categories.*

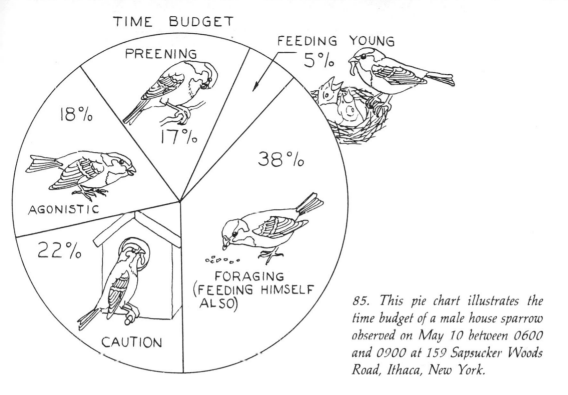

TIME BUDGET

PREENING

FEEDING YOUNG
— 5%

18%

17%

38%

AGONISTIC

22%

FORAGING
(FEEDING HIMSELF
ALSO)

CAUTION

85. This pie chart illustrates the time budget of a male house sparrow observed on May 10 between 0600 and 0900 at 159 Sapsucker Woods Road, Ithaca, New York.

To represent the time budget on a pie chart, calculate angles by multiplying 360 degrees by the percentage of time for each behavior category. Draw a circle with a compass and measure angles with a protractor. Sketches of typical behaviors lend a nice touch if you feel artistic.

Although ethograms, behavior sequences, and time budgets may be similar for different individuals of the same species, there will also be individual variations arising from different personalities. It takes a keen field observer to recognize individual personalities, but this offers an added challenge that gives behavior watching infinite appeal. Focus your behavior watching with these points in mind:

1. Record circumstances surrounding your observations. Include location, time of day, weather conditions, presence of other animals.
2. Quantify observations whenever possible by noting how long, how many times, or in what sequence behaviors occur.
3. Minimize your presence by using binoculars and by watching from a blind or other cover. Consider your effect(s) on the behavior you are watching.
4. Keep interpretations separate from descriptions.
5. Avoid assigning human motives (birds are not little people dressed in feathers).
6. Assign names to specific behaviors and then describe the behaviors in detail.
7. Be patient, especially if you are watching during the "off hours," between 11 A.M. and 3 P.M., when birds are less active.

Publishing Your Observations

If you observe an unusual behavior or conduct a thorough behavioral study, research your topic in the bird literature and discuss your observations with professionals at a local museum or university (see Chapter 5 for a list of ornithology contacts). If, after checking these resources, you believe that your observations are unique, study appropriate formats and write a short note or paper for one of the bird journals described in Chapter 7.

TAKING FIELD NOTES

Record your bird observations carefully, for even the most casual excursions can produce useful scientific data. A small investment of time taking notes in the field and later organizing them at home can produce a valuable set of field notes. Such notes will have increasing value if they can be clearly interpreted by others and if information is available for quick retrieval. The sections that follow are a few suggestions for improving your note-taking skills and for filing observations so they can be found easily when needed.

Checklists

Checklists are the most frequently used type of field record, but, unfortunately, most lists contribute little to our understanding of bird distributions and abundance. Most observers lazily check off the species they have seen without indicating the number seen or providing details about where the birds were observed. In the search for new and unusual species, many birders ignore the most abundant birds. Such records would be much more valuable if they included an actual count or an estimate of numbers for *all* species.

In addition to indicating numbers, always note the exact location (direction in air miles from nearest town), hours in the field, and weather conditions. Most checklists provide space for these items. With these basics, your checklist will have meaning for anyone who might be interested in using the data in the future. Complete, personal, daily checklists are useful for preparing and updating local and regional checklists, and their value increases with time because such counts provide important baseline data for detecting population changes. Information from a number of observers over a broad area can serve as an early warning signal that a formerly abundant species is experiencing a serious population decline.

Field Notebooks

While in the field keep a pocket notebook for recording observations. When describing field marks and behaviors of unknown or unusual birds, include details of color, plumage, and behavior as well as observation conditions, such as lighting, approximate distance from the birds, and names of people who saw the birds. Thorough, written descriptions by several people will help to verify unknown birds at a later time and are often accepted by authorities as the basis for legitimate sight records.

Microcassette tape recorders are a convenient alternative to field notebooks. Their pocket size and battery efficiency make them ideal field tools. The digital counter feature is a great aid in locating notes and is well worth the extra few dollars. After you transcribe the data at home, you can reuse the tape cassettes many times.

The Birder's Field Notebook, edited by Susan Roney Drennan, offers a convenient way to record details about unfamiliar birds. This pocket notebook contains a standard two-page format for recording field descriptions of seventy-two different birds. It is important to observe unknown birds systematically for all field marks and behaviors necessary to verify field observations. Record these at the time of the observations so that details are not forgotten. A good technique to record field marks is to start by looking at the bird's head and work your way back to the tail. With practice, a quick sketch can capture behaviors and map your descriptions of field marks.

FIELD IDENTIFICATION REPORT

Date: _____ Habitat: _____

Location: _____ , _____
     (Distance from nearest city)   *(State/Region)*

Weather: Temperature _____ , Wind _____ , % Cloud Cover _____
                      (Direction/Speed)

Observers: Binoculars   Spotting Scope
                               (Circle One)

Distance from Bird:

Field Marks:

Behavior:

Vocalizations:

Comments (Sketch):

Record data for field identification reports
at the time of the observation rather than
attempting to recall details upon returning home.

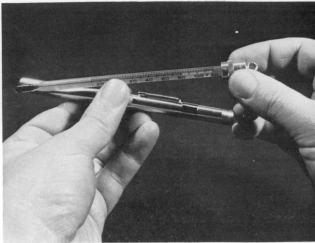

86. For only a few dollars, you can obtain inexpensive tools for measuring temperature and wind speed. Measure wind speed with a Dwyer Wind Meter (left), and measure temperature with a Pocket Case Thermometer (right). Both are available from Forestry Suppliers, Inc., Box 8397, 205 Rankin Street, Jackson, Miss. 39204.

The Grinnell Field Note System

Because of its straightforward approach and the need for a standardized note-taking scheme, the Grinnell Field Note System has been widely accepted. Established by Joseph Grinnell of the Museum of Vertebrate Zoology, University of California at Berkeley, the system is based on three record-keeping sections: the Catalog, Species Accounts, and the Journal. Grinnell's Catalog is widely used by museum workers to list collected specimens with details about locality, measurements, and other details. His system of recording Species Accounts and Journal entries has a much broader application and is the standard tool of many field observers. (Joseph Grinnell insisted on recording field observations directly into Species Accounts and Journal notebooks while in the field to avoid loss of detail and errors resulting from transposing. However, a pocket field notebook or tape recorder provides a convenient way to record details under rigorous field conditions and while traveling.)

SPECIES ACCOUNTS

Soon after returning home from the field, transfer observations from your field notebook to individual Species Accounts. In this system, observations about a particular species are grouped together, either as lists or sightings or as more detailed behavioral or identification accounts. The great benefit of the system is that all observations about a species are grouped together, rather than spread throughout your field notes or Journal. In this system, the Species Accounts appear on loose-leaf pages (the preferred size is approximately 6½ × 9 inches). The loose-leaf feature permits you to add additional pages to existing notes on each species.

Add observations of special interest to a list of running accounts, giving date, exact locality, number of individuals, nests, eggs, and any details about age, sex, or behavior. Keep running accounts only for selected species that are of personal interest. The system is especially useful for recording unusual or declining species (see Blue List Project, page 183).

When you see an unusual or unfamiliar bird, take abundant field notes, describing all details, and add this information to your Species Accounts. Detailed behavioral accounts should also be transferred from your field notebook (or tape recorder) to a page in your Species Accounts notebook. Species Accounts need not be limited to unusual birds. Pick a common bird that interests you and start assembling Species Accounts. It won't be long before you become a local "authority" on the species. To keep your Species Accounts organized, arrange them to follow the taxonomic order presented on a field checklist (see illustrations 87 and 88).

JOURNAL

The Journal is a daily summary of your itinerary and observations, with supporting details about weather, names of field companions, and places visited (see illustration 89). Most observers prefer loose-leaf notebooks, but bound Journals are also adequate. The Journal provides an excellent place to tally the species you've seen during the day by transferring lists and numbers from your field notebook, tape recorder, or checklist. Underline unusual species or numbers and note which birds are detailed further in your Species Accounts notebook. Whenever possible use abbreviations, but be certain to include a key so there is no doubt about the meaning of your notations. The following are a few standard Journal abbreviations:

Sp. Acct.	=	Species Account
♂ ,♂♂	=	male, males
♀ ,♀♀	=	female, females
imm.	=	immature
N/C	=	nest under construction
N4E	=	nest with four eggs
N3Y	=	nest with three young
A/F	=	adult carrying food
H.O.	=	heard only
D.O.R.	=	dead on road
F.O.	=	flying over
~	=	approximately (for example, ~100 = approx. 100)

Also include comments about unusual trends and events of the day, but do *not* enter detailed accounts of behavior and descriptions of unusual species; they are easily lost in your Journal entries and are easier to retrieve from your Species Accounts.

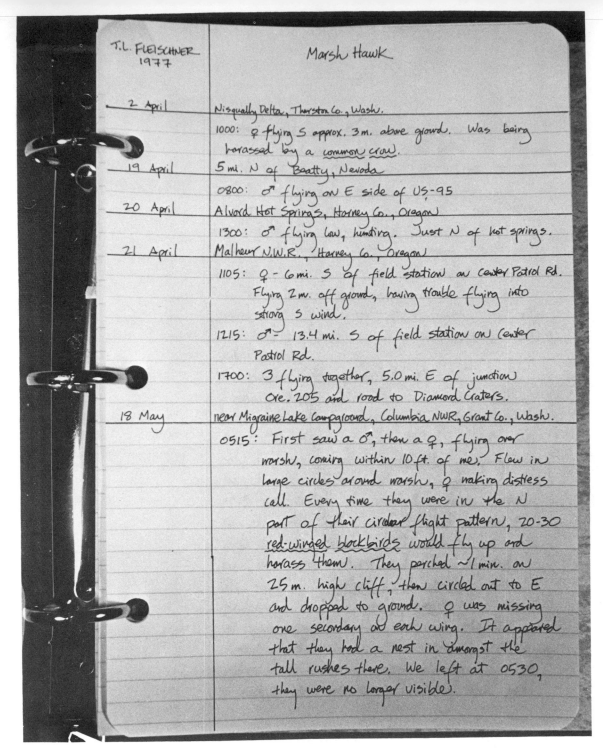

T.L. FLEISCHNER 1977	Marsh Hawk
2 April	<u>Nisqually Delta, Thurston Co., Wash.</u>
	1000: ♀ flying S approx. 3m. above ground. Was being harassed by a common crow.
19 April	<u>5 mi. N of Beatty, Nevada</u>
	0800: ♂ flying on E side of US-95
20 April	<u>Alvord Hot Springs, Harney Co., Oregon</u>
	1300: ♂ flying low, hunting. Just N of hot springs.
21 April	<u>Malheur N.W.R., Harney Co., Oregon</u>
	1105: ♀ - 6 mi. S of field station on Center Patrol Rd. Flying 2m. off ground, having trouble flying into strong S wind.
	1215: ♂ - 13.4 mi. S of field station on Center Patrol Rd.
	1700: 3 flying together, 5.0 mi. E of junction Ore. 205 and road to Diamond Craters.
18 May	<u>near Migraine Lake Campground, Columbia NWR, Grant Co., Wash.</u>
	0515: First saw a ♂, then a ♀, flying over marsh, coming within 10 ft. of me. Flew in large circles around marsh, ♀ making distress call. Every time they were in the N part of their circular flight pattern, 20-30 <u>red-winged blackbirds</u> would fly up and harass them. They perched ~1 min. on 25 m. high cliff, then circled out to E and dropped to ground. ♀ was missing one secondary on each wing. It appeared that they had a nest in amongst the tall rushes there. We left at 0530, they were no longer visible.

87. Following the format of this sample page from a Species Account on the marsh hawk, all sightings for a species should be listed on the same page. Additional loose-leaf pages are added when necessary. Localities should be underlined, and these names should correspond to Journal entries for the same date.

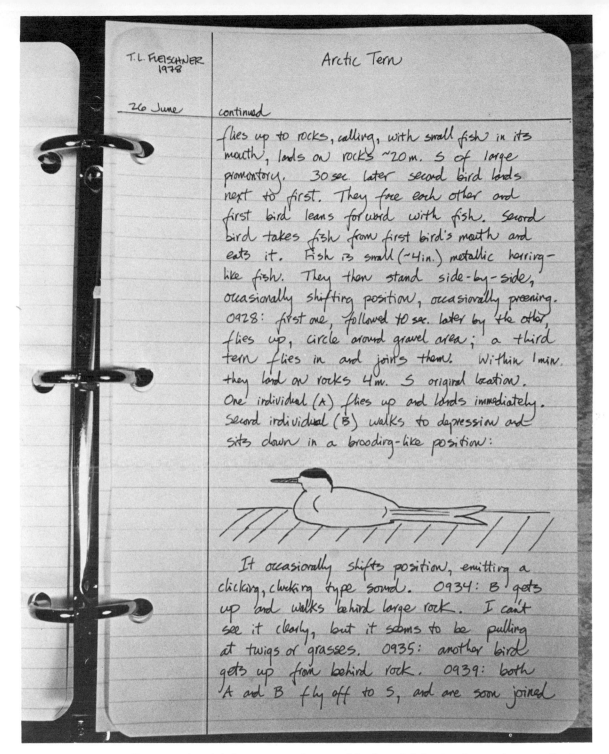

T. L. Fleischner
1978

Arctic Tern

26 June

continued

flies up to rocks, calling, with small fish in its mouth, lands on rocks ~20m. S of large promontory. 30 sec. later second bird lands next to first. They face each other and first bird leans forward with fish. Second bird takes fish from first bird's mouth and eats it. Fish is small (~4in.) metallic herring-like fish. They then stand side-by-side, occasionally shifting position, occasionally preening. 0928: first one, followed ½ sec. later by the other, flies up, circle around gravel area; a third tern flies in and joins them. Within 1 min. they land on rocks 4m. S original location. One individual (A) flies up and lands immediately. Second individual (B) walks to depression and sits down in a brooding-like position:

It occasionally shifts position, emitting a clicking, clucking type sound. 0934: B gets up and walks behind large rock. I can't see it clearly, but it seems to be pulling at twigs or grasses. 0935: another bird gets up from behind rock. 0939: both A and B fly off to S, and are soon joined

88. Organize detailed descriptions of unusual birds and behavioral notes in your Species Account notebook, as shown in this sample page on the Arctic tern. Underline localities at the beginning of each entry for quick reference to the Journal entry of the same date.

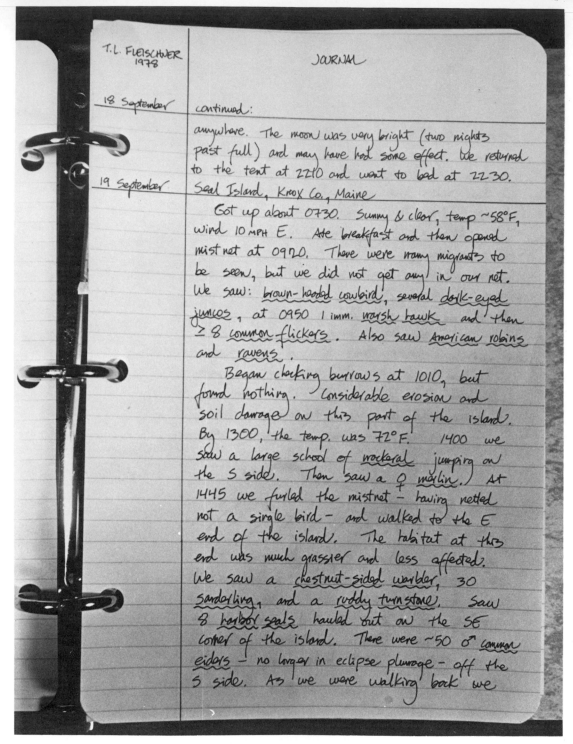

T. L. FLEISCHNER 1978	JOURNAL

18 September | continued:

anywhere. The moon was very bright (two nights past full) and may have had some effect. We returned to the tent at 2210 and went to bed at 2230.

19 September | Seal Island, Knox Co., Maine

Got up about 0730. Sunny & clear, temp ~58°F, wind 10 MPH E. Ate breakfast and then opened mist net at 0920. There were many migrants to be seen, but we did not get any in our net. We saw: brown-hooded cowbird, several dark-eyed juncos, at 0950 1 imm. marsh hawk and then ≥ 8 common flickers. Also saw American robins and ravens.

Began checking burrows at 1010, but found nothing. Considerable erosion and soil damage on this part of the island. By 1300, the temp. was 72°F. 1400 we saw a large school of mackeral jumping on the S side. Then saw a ♀ marlin. At 1445 we furled the mistnet — having netted not a single bird — and walked to the E end of the island. The habitat at this end was much grassier and less affected. We saw a chestnut-sided warbler, 30 sanderling, and a ruddy turnstone. Saw 8 harbor seals hauled out on the SE corner of the island. There were ~50 ♂ common eiders — no longer in eclipse plumage — off the S side. As we were walking back we

89. As this sample page shows, the Journal is a place to summarize your day's observations and to comment in general terms about your field observations.

GUIDELINES FOR WRITING JOURNAL AND SPECIES ACCOUNTS ENTRIES

1. With a straightedge, draw a 1½-inch margin on the left side of each page and center either the word JOURNAL or the name of the species concerned (for Species Accounts). Then place your name and the year in the upper left-hand corner.

2. For Journal entries include your name, year, date, locality, route of travel, weather, habitat description, commentary, and species list. For Species Accounts include your name, year, date, locality, and commentary.

3. All information is recorded consecutively. Don't begin a new page for each new entry, but each page should be capable of standing alone. Repeat your name, year, date, and center heading on each page, noting "continued," if an entry extends from one page to the next.

4. Write neatly, using concise sentences.

5. Reduce wasted space by extending sentences all the way to the right-hand edge of the page and listing species in running form.

6. Write on only one side of a page. This reduces "bleed through" problems with copy machines and leaves the opposite page available for supplementary maps and sketches.

7. Underline species names with wavy lines so that they will be easy to locate at a glance. For example: hedge mustard, song sparrow.

8. Use the international 24-hour clock system to report times.

9. Use standard abbreviations and avoid slang expressions.

10. Give top priority to compiling Journal entries at the end of each day from your field notes. If you fall behind, do the current day before going back to catch up on previous days.

For a more detailed discussion of the Grinnell Field Note System with abundant sample pages, consult *The Naturalist's Field Journal: A Manual of Instruction Based on a System Established by Joseph Grinnell* (see "Building a Bird-Watcher's Library," page 244).

Note-taking Equipment

The proper choice of paper and ink will greatly enhance the life of notes and their legibility in years to come. Species Accounts and Journal entries are best recorded with a technical pen and a quick-drying waterproof black ink. There are several popular name brands of technical pens, such as Rapidograph and Staedtler. The pens come with various size tips, but the preferred width of the line for note taking is approximately 0.35mm. Pens that draw a narrower line are more likely to clog. Frequent use is the best guarantee that your pen will be ready when you need it and not be clogged by dry ink. When traveling, wrap the technical pen in tissues and place it in a small plastic bag to absorb leaks that might be caused by vibrations and altitude changes.

Select a paper with a high rag content. It will hold the ink better and will not turn yellow and brittle with age.

Preserving Your Field Notes

It doesn't take long for serious field observers to collect a small mountain of notebooks filled with checklists, Species Accounts, and Journals. If the data is well prepared, it will become increasingly valuable and, therefore, deserves protection. A duplicate copy of the notes or storage in a fireproof safe are important safety considerations. For a permanent repository, check with a local or state vertebrate museum, showing them a sample of your notes, and make arrangements in your will so that the notes are properly preserved. Do not underestimate the value of your field records.

Punch Cards

Data punch cards may be the best system to use if you are collecting large amounts of data and need quick retrieval. In addition to providing a flexible system of coded data, some cards also provide space for written notes in the center of each card. To order a special newsletter that clearly details how amateur bird-watchers can use punch cards, write Indecks, Inc., Arlington, Vt. 05250, to request a copy of "Getting the Most Out of Backyard Birding."

COUNTING BIRDS

The ability to count birds accurately is a skill that comes only with considerable practice. It becomes increasingly difficult to make exact counts when birds occur in large numbers or where several species are present together. But even difficult, laborious counts are more useful than "ballpark estimates." The challenge is to give as accurate a count as possible.

Depending on such variations as lighting and distance to the birds, it is best to present round numbers when exact counts are not possible. A flock containing 100 or fewer individuals should be rounded off to the nearest 5 or 10 birds; a flock with more than 100 birds may be rounded to the nearest 25 or 50, depending on viewing opportunities. Avoid giving range estimates of birds, such as "approximately 300 to 500," since such estimates are unnecessarily vague and are difficult to compare with other counts.

Flocks of flying birds, such as waterfowl, shorebirds, and blackbirds, are among the most difficult to count. Their speed, movement, and habit of flying in dense three-dimensional flocks contribute to the difficulty of achieving reasonable estimates. There are a few techniques that will help.

If you are attempting to determine the number of birds congregating at a certain point (such as herons or blackbirds returning to a roost), sample the number passing a fixed point (such as a tree or house) in one-minute intervals throughout the period during which the birds return. Average your one-minute counts and multiply the average by sixty to find the number of birds passing the reference point in one hour. For a more complete estimate of birds moving past you to the roost, determine the total duration of the procession from start to finish, then multiply this time period (in minutes) by the number of birds seen in an average minute.

Exact counts are usually possible if there are fewer than thirty birds in a flying flock. For larger numbers try a technique called blocking. This approach consists simply of counting a block of birds of typical density from the trailing end of the flock and then visually superimposing this block onto the rest of the flock to see how many times it will fit. For example, if a flock contains about sixty birds, your block need only contain the trailing twenty birds in the group, and this would "fit" onto the remainder of the flock two more times.

For huge flocks, start by estimating part of the group to represent a block of 50 or 100 birds. Then use this sample to arrive at a total estimate, just as with smaller numbers. Concentrate on memorizing impressions of what different size flocks look like. With practice you can develop mental images because different groupings have distinctive shapes and forms.

It is surprising how varied estimates can be from different observers who are all looking at the same birds. Skill at estimating bird numbers depends largely on practice. A good way to sharpen estimating accuracy is to make a set of flash cards with different numbers of dots. Mark the number of dots on the back of the cards and then have a friend check your counting ability by flashing the cards at various moving speeds. A similar system consists of having a friend toss precounted samples of beans or popcorn onto a tabletop. Practice making quick estimates until you achieve accuracy within 5 percent.

You can also practice your estimating skill by making projection slides with different numbers of dots. Project the slides at a bird club meeting and compare es-

90. The blocking technique permits estimates of large flocks. Count the number of birds in a sample block and then see how many additional times this block can be superimposed on the remainder of the flock.

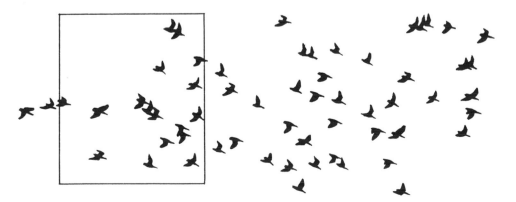

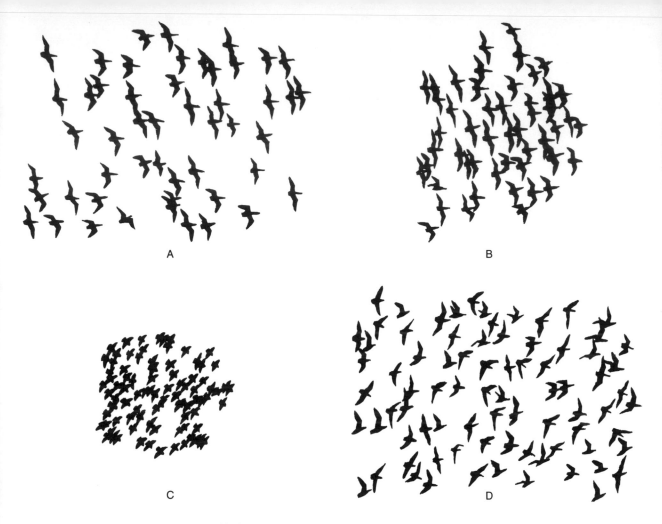

91. Practice the blocking technique of counting birds with these four flock patterns. The number of birds in each flock is given in a footnote on page 74.

timates. Such slides are easily constructed by poking pinholes into 2 × 2-inch pieces of cardboard or opaque color slides. Challenge club members to make estimates by keeping the "flocks" on the screen for only a few seconds. This exercise is especially useful in training group members to participate in such counts as the Audubon Christmas Bird Count.

Many North American bird censuses and surveys rely on amateur participants (see Chapter 6). The success of these important studies depends largely upon the counting skills of participants. Practice the blocking technique by estimating the number of birds in illustrations 91A–D. The answers are given in a footnote* on page 74.

SKETCHING BIRDS

Any careful observer can make useful field sketches. Regardless of artistic quality, field sketches serve as valuable visual records of the birds you encounter. Start by sketching in the field and you'll be surprised at the additional details of form and behavior you begin to notice. Look at the shape and posture of the body, paying special attention to the relative proportions of head, wings, and tail. Consider bird anatomy as you sketch, carefully noting positions of head, wings, tail, and legs. The following illustrations will help you with basic bird anatomy.

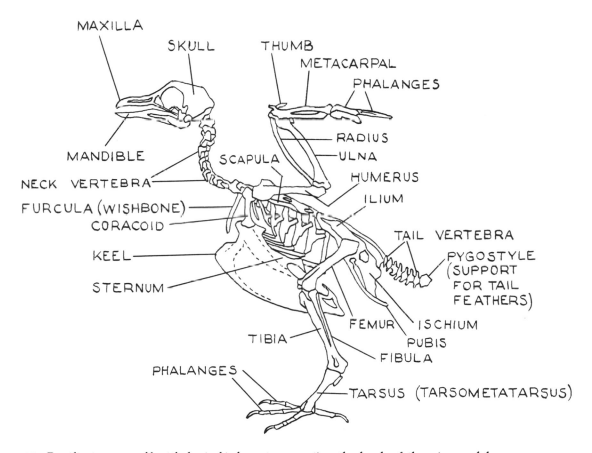

92. Familiarize yourself with basic bird anatomy, noting the bends of the wing and leg and how these limbs attach to the avian skeleton. Adaptations for flight have affected bird anatomy in many ways, but similarities to the human skeleton remain obvious.
Illustration redrawn from *Ornithology in Laboratory and Field* by Olin S. Pettingill, Jr., with permission of Burgess Publishing Company

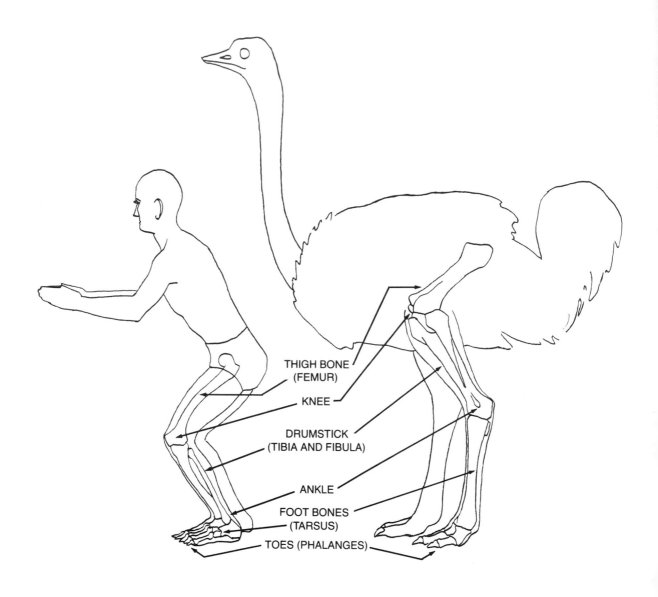

THIGH BONE
(FEMUR)

KNEE

DRUMSTICK
(TIBIA AND FIBULA)

ANKLE

FOOT BONES
(TARSUS)

TOES (PHALANGES)

93. Note that the avian thigh bone (femur) is shortened and buried in the bird's body with the forward end lying near the bird's center of gravity. The drumstick bone (tibia and fibula) emerges from the bird's body ending in the ankle. Foot bones are fused into one bone called the tarsus, which points forward, ending back at the bird's center of gravity. Birds walk, hop, stand, and run on their toes.

* The number of birds in illustrations 91A–D are: (A) 50, (B) 50, (C) 95, (D) 80.

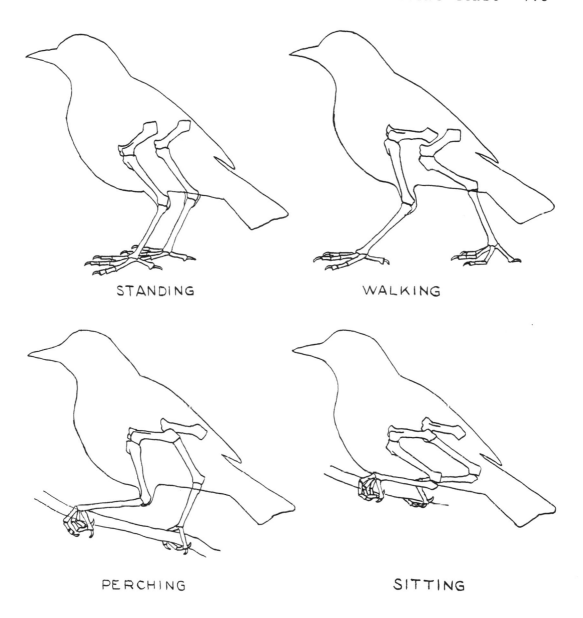

STANDING

WALKING

PERCHING

SITTING

94. These four illustrations show a perching bird leg in various postures and the position of the toes and leg bones as the bird stands, walks, perches, and sits.

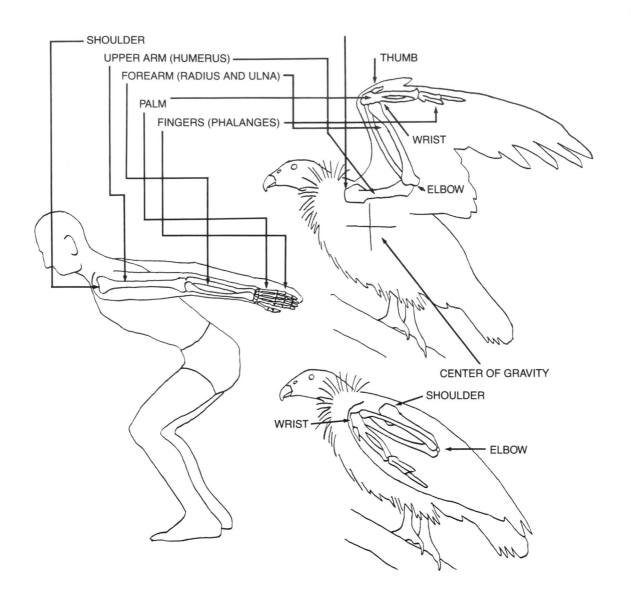

SHOULDER

UPPER ARM (HUMERUS)

FOREARM (RADIUS AND ULNA)

PALM

FINGERS (PHALANGES)

THUMB

WRIST

ELBOW

CENTER OF GRAVITY

SHOULDER

WRIST

ELBOW

95. *Bird wings are attached above and slightly forward of the center of gravity. When a bird unfolds its wings over its back, the upper arm (humerus) points backward to the elbow, where it connects to the forearm (radius and ulna), which bends forward ending at the wrist. Elongated and fused hand bones point back toward the tail. Wing bones compactly fold together when the bird is at rest.*

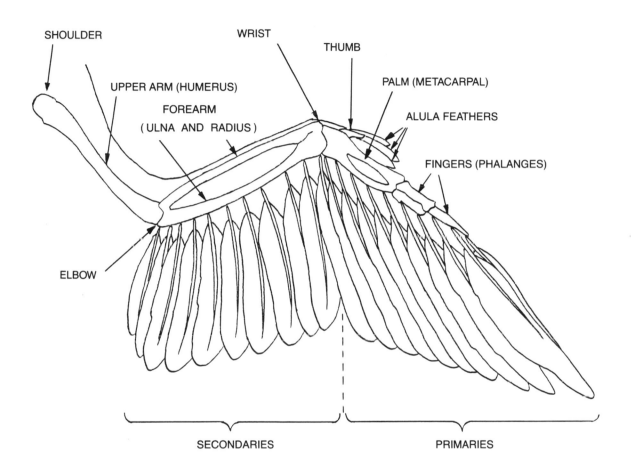

SHOULDER

WRIST

THUMB

UPPER ARM (HUMERUS)

PALM (METACARPAL)

FOREARM
(ULNA AND RADIUS)

ALULA FEATHERS

FINGERS (PHALANGES)

ELBOW

SECONDARIES

PRIMARIES

96. Secondary feathers are firmly attached to the broad ulna of the forearm. The large flight feathers (primaries) attach to the modified palm and finger bones. The number of primaries varies from as few as nine in small land birds to as many as twelve in large birds, such as storks and flamingos. Many hummingbirds have only six secondaries, but the long-winged wandering albatross has thirty-two. The tiny thumb supports several small feathers called the alula.

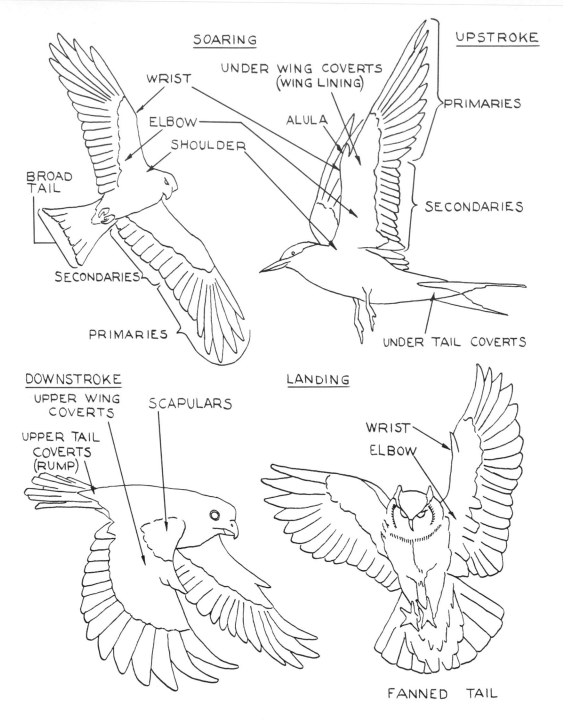

SOARING

WRIST

UNDER WING COVERTS
(WING LINING)

ELBOW

ALULA

SHOULDER

BROAD
TAIL

SECONDARIES

PRIMARIES

UPSTROKE

PRIMARIES

SECONDARIES

UNDER TAIL COVERTS

DOWNSTROKE

UPPER WING
COVERTS

SCAPULARS

UPPER TAIL
COVERTS
(RUMP)

LANDING

WRIST

ELBOW

FANNED TAIL

97. In these common flight postures—soaring, upstroke, downstroke, and landing—note the positions of the wings and how birds variously use their tails as a balance rudder or brake to assist wing action.

Assembling a Field Sketch

Start your field sketch with an oval that approximates the general proportions of the bird. Regardless of whether you intend to sketch an owl, heron, or robin, they all have oval (egg-shaped) bodies (see illustration 98). It is the differences in wings, tails, and legs that give each species a distinctive form. Watch carefully to

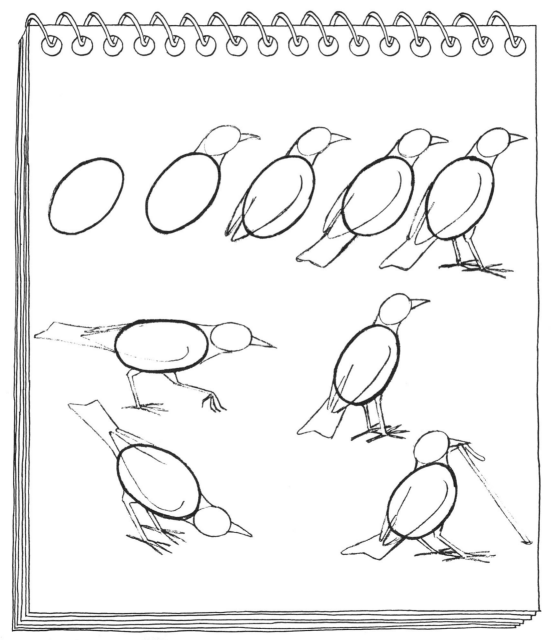

98. Building a robin sketch and sketches of behavior.

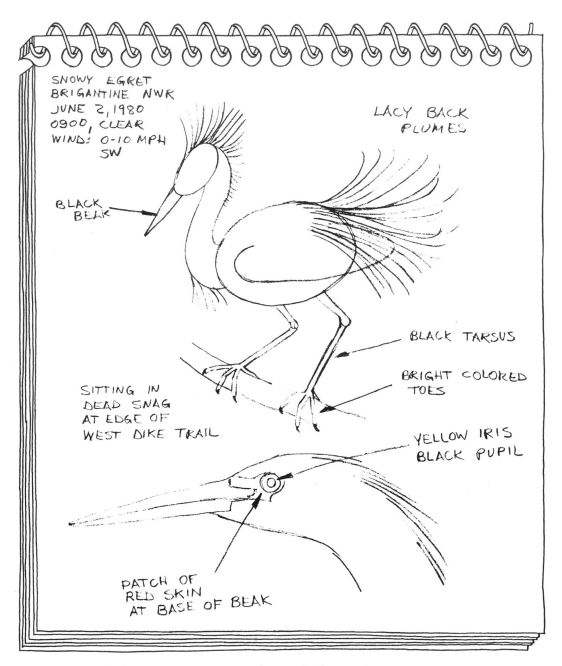

99. *Detailed field sketch of a snowy egret with notes in the margin.*

see at what angle the bird holds its body, then begin to assemble body parts, out-lining head and neck, wings, tail, and legs. Concern yourself with proportions of the different parts to one another and the position of attachment, always referring to the living bird.

Draw with smooth, flowing lines to achieve an outline sketch of the bird. Don't worry about erasing mistakes or lines you don't like. The goal should be to cap-ture shape and posture with as few lines as possible before the bird flies away.

Portray most behaviors by changing the posture of the bird's body (position of the oval) and the position of its appendages. If you see an unusual bird or one you can't identify, quickly draw a standard perching posture and then add details to illustrate distinctive field marks, carefully noting these in the margins of the page (see illustration 99).

Sketch birds in your field notebook or in a small artist's spiral sketchbook using a soft HB pencil, adding finer details with a harder pencil, such as a 4H. An art-gum eraser is useful for cleaning up sketches, but erasures should be kept to a minimum. After returning home, transfer sketches to your Species Accounts note-book or Journal.

Practice by sketching tame birds, such as captive parakeets, pigeons, or feeder birds. Sitting postures are easiest, but it won't take long before a few pencil lines will also capture the movement of birds in flight.

4

PHOTOGRAPHING AND RECORDING BIRDS

Photography permits a longer, slower, and more understanding view of birds. It can sharpen observation skills and provide opportunities to record intimate details of bird behavior that might otherwise go unnoticed. Even poor-quality photographs can document the occurrence of rare birds, a fact that has helped to retire more than a few museum collectors' shotguns. For most birders, however, the greatest satisfaction gained from bird photography comes from sharing favorite photos with others and vividly recalling memorable bird-watching experiences. Recorded bird sounds will enhance any slide program you present to share these experiences.

SELECTING PHOTOGRAPHIC EQUIPMENT

Cameras

Most bird photographers prefer the 35mm single-lens-reflex format. The principal advantage of this design is that it permits the photographer to look through the viewfinder and see exactly what will be in the photograph.

Because 35mm cameras are lightweight and compact, they are ideal for the birder who is already encumbered with binoculars and field guides. Another important feature is the ease with which lenses can be changed, providing flexibility to photograph everything from close-ups of flowers to telephoto views of birds.

Although there is widespread enthusiasm for single-lens-reflex cameras among bird photographers, selection of a camera is complicated by opinionated disputes about the various merits of different models. Different opinions about durability and the comparative value of various features, such as manual versus automatic exposure systems and aperture versus shutter-preferred mechanisms, can make selection of the "right" camera a bewildering experience.

Do not let debates about equipment get in the way of selecting a camera. Any of the better 35mm single-lens-reflex models are suitable for taking excellent bird photographs. Ask for advice from friends who take successful pictures, then purchase the best camera you can afford. If you select the same camera model as an experienced local photographer, you are likely to have a good source of advice as you develop your own photographic style.

Remember that even automatic cameras cannot take good pictures by themselves. Only an experienced and sensitive photographer can recognize a photographic opportunity, compose the picture, consider the light, and snap the shutter at the right time. Expensive equipment in itself is no guarantee that satisfying photographs will result.

Telephoto Lenses

MAGNIFICATION

The magnifying power of a telephoto lens is indicated by the focal length of the lens. Magnification increases by one power for every 50mm increase in focal length. For example, a 100mm telephoto lens makes the subject look twice as close as it would through a standard (normal) 50mm lens. Likewise, a 400mm telephoto would magnify a subject eight times and be equivalent to an 8x binocular.

Just as it is difficult to hold high-power binoculars steady, it is also difficult to hand-hold a long telephoto lens—especially if you are attempting to photograph birds from a rocking boat. Usually it is inadvisable to hand-hold a telephoto larger than 300mm. Steady such telephotos with a tripod or other support.

CLOSE-FOCUSING CAPABILITY

To fill the camera field of view, it is important to focus at close distances to the subject. Close-focusing capability is especially useful for small birds, for even with a 400mm lens one must approach surprisingly close to take an interesting picture. For birds the size of a robin, the lens must focus to approximately 15 feet to fill the field; using the same 400mm telephoto, a warbler would only fill the field at 8 feet from the lens. As a general rule, avoid telephotos that will not focus closer than 15 feet, but be ready to pay more for those with close-focusing capabilities.

EXTENSION TUBES

Extension tubes offer a way to help fill the field with bird images. They serve as spacers, permitting the lens to move farther from the film surface. They contain no glass and, therefore, do not affect image sharpness. Aside from minor loss of light, the only drawbacks to using extension tubes are the possibilities of dark photo borders (vignetting) and less focusing flexibility should the bird move out of close range.

LENS SPEED

The speed of the lens refers to the maximum aperture or diaphragm opening available. Aperture size is expressed as f numbers; the smaller the f number, the larger the aperture opening. Telephotos with larger maximum apertures permit the use of faster shutter speeds, which reduces vibrations and results in sharper pictures. For example, a lens with a maximum aperture of f/8 might permit shooting at $\frac{1}{125}$ second, but a lens with a maximum aperture of f/5.6 would, under the same conditions, permit shooting at $\frac{1}{250}$ second and result in a sharper picture. Fast lenses also brighten viewing fields, and this makes it easier to focus accurately and quickly. However, fast lenses have correspondingly wider diameter objective lenses, and the additional glass makes them both heavier and more expensive. Within limits, these are generally worthwhile trade-offs for the sharper pictures that result. For these reasons it is best to avoid telephotos that are slower than f/8.

SHUTTER SPEEDS

Large telephoto lenses require fast shutter speeds to reduce vibrations and the chance of blurred images. Using a medium-speed film (such as ASA 64), a shutter speed of $\frac{1}{60}$ second is usually adequate for lenses with focal lengths of 100mm or less. Longer telephotos require faster shutter speeds. For comparison, the minimum shutter speed for lenses in the 100–300mm range should be $\frac{1}{125}$ second. If light permits, larger lenses require even faster shutter speeds. If an unusual bird appears or an interesting behavior occurs, it is well worth the cost of film to brace your lens, hold your breath, and take a chance at a slower shutter speed. With practice and a critical review of processed photos, you will soon learn how steady you can hold your various lenses.

DEPTH OF FIELD

Telephoto lenses usually produce photos with a characteristically shallow depth of field (the depth of the picture that is in focus). Depth of field decreases as magnification (focal length) and aperture size increase. Depth is further reduced by focusing on close subjects. For example, a 400mm lens opened to f/5.6 and focused at 15 feet has a depth of field of only one inch!

To increase the amount of depth in your pictures, "stop down" the lens diaphragm to a smaller opening (larger f number), but remember that this means shooting at a slower shutter speed and risking blurred pictures from too much movement.

ALL-GLASS VERSUS MIRROR TELEPHOTOS

There are two basic telephoto lens designs: all-glass and mirror. All-glass telephoto lenses offer the sharpest images, brightest fields of view, and best color rendition. Their principal drawbacks are bulky size and heavy weight, which make them difficult to carry in the field and hold steady. Mirror lenses are comparatively easy to carry in the field because of their lighter weight and compact design. A further advantage is the excellent close-focusing capability that gives mirror lenses a distinct advantage over all-glass telephotos. However, the mirror design may produce color shift, light fall-off at the border of the picture, and donut-shaped patterns in the background. Another drawback is the slow, fixed aperture (usually f/8), which makes mirror lenses harder to focus and difficult to use in low-light situations.

AUTOMATIC VERSUS PRESET LENSES

If you choose an all-glass telephoto, you must then select either an automatic or a preset design. The diaphragm of an automatic lens can be set at the preferred opening, but it will not actually close down until the shutter is released. This permits focusing with the lens wide open, a feature that facilitates sharper and quicker focusing. In contrast, preset lenses are slower to use because the lens diaphragm must be closed manually to a preselected opening after focusing with the diaphragm wide open. For this reason, automatic lenses are easier and quicker to focus than preset lenses, though they are more expensive than preset models. Focusing considerations are especially important for bird photography, because telephoto lenses have a shallow depth of field, and the active nature of birds usually necessitates frequent focusing adjustments.

ZOOM LENSES

Lenses with zoom focal-length combinations such as 70–210mm and 100–300mm offer a convenient way to photograph birds at varying distances with a minimum amount of lens juggling. Macrozooms offer added convenience by doubling as a close-up lens, permitting the photographer to focus on everything from the distant horizon to subjects within a few inches of the lens. However, beware of the lower-priced zoom lenses, because they are likely to produce less-than-crisp images and may suffer from loss of light at the edges of the field.

TELECONVERTERS

Teleconverters fit between the lens and camera body. They increase the magnification of telephoto lenses and are available in several powers: 1.5x, 2x, 3x, and 4x. For example, a 2x teleconverter mounted to a 300mm lens will increase magnification to the equivalent of a 600mm telephoto. Although some manufacturers match teleconverters to individual lenses with success, most converters offer disappointing results because they disrupt resolution, reduce available light, and make it difficult to hold the lens steady. The larger the magnification of an "extender," the greater the negative effects.

100. Zoom lenses are the convenient answer to most short telephoto needs.
Photo by Michael J. Hopiak

TELEPHOTO SUPPORTS

Shoulder supports offer a convenient way to carry your big lens and camera in the field, but for lenses that are longer than 500mm, a secure tripod is the best answer. Small sandbags offer another option. If you are photographing birds from a car, place the bag on the car roof and nestle your big lens into the bag to gain support. If you must hand-hold a long telephoto, cradle the large end of the lens in one hand and brace your upper arm against your chest.

Flash Equipment

Flash equipment can make it possible to photograph birds on overcast days and to light adequately such dark habitats as forests and thickets. Flash equipment is most useful for moderately close-up work, such as photographing birds at nests and feeders.

Features to consider when selecting a flash unit are brightness, duration of flash, the time it takes between exposures for the light to become ready for reuse (recycling interval), and the bulk and weight of the unit. Flash units are rated for brightness by their guide numbers; the larger the guide number, the brighter the flash. The duration of the flash should be at least $1/1000$ second to stop bird move-

101. Telephoto supports are a must for lenses larger than 300 mm.
Photo by Michael J. Hopiak

ments. Flashes with durations of 1/15,000 second are necessary to stop the wings of hummingbirds.

Inexpensive flash equipment with a guide number of approximately 65 is adequate for photographing birds within about 3 feet of the flash unit. To photograph birds at a nest or feeder, mount the flash unit on a tripod and position it approximately 3 feet from where the birds are likely to perch. Set your camera shutter to the special "x" synchronization speed, and calculate your aperture setting by dividing the flash unit's guide number by the distance to the subject (with experience you'll arrive at correction adjustments for your particular flash unit).

Manual flash systems are adequate for nest and feeder photography, because the flash is mounted on a tripod, and the distance to the bird is predictable. Shoot a test roll to establish the correct settings with your flash set at carefully measured distances from the subject. Don't trust the dial on the back of your flash unit. Once established, the f number will always be the same, and you won't have to worry about proper exposure. Then you'll be ready to capture the action when you see it.

Special automatic flash units designed to produce a concentrated light for use with telephoto lenses are useful for stalking up to a bird where distance is not known and speed is critical. However, always take care that automatic units are not fooled by unusual lighting situations. If you suspect that your light unit may be fooled by backlighting, a white bird, or some other unusual situation, bracket the predicted f setting by also exposing frames at one-half and one full stop below and above to improve your chances of obtaining a satisfying photo.

A second flash head, called a slave unit, can reduce distracting shadows produced by overhead lighting. Slave units are best synchronized to flash simultaneously with the master flash head by an extension cable connecting the slave unit to the camera (some flash units have a power cable for each head connecting directly to a portable battery pack). Mount the slave unit on a tripod and direct the light at the bird's perch from the front and to the side to provide fill light that washes out shadows from the master flash (see illustration 102).

102. Flash A (master) provides light from above similar to the way sunlight would fall on the bird. Flash B (slave) is a fill light that provides light from the front and eliminates distracting shadows from Flash A.

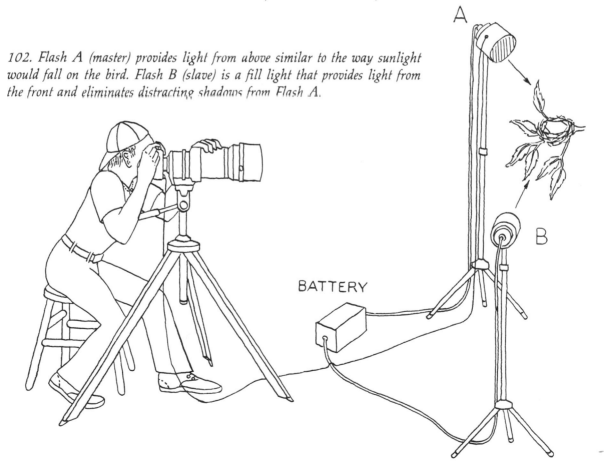

FILM, PROCESSING, AND HOME PRINTING

Color films are available either as transparency (slide) film or color print film. Most bird photographers prefer color slide film because it is less expensive per frame and because publishers prefer slides to color prints. A further advantage to using slide film is that while both color and black-and-white prints are easily made from transparencies, it is much more expensive to make slides from prints, and the results are less satisfying.

Regardless of the manufacturer, slide films are easily recognized by the word "chrome" incorporated into the film name, e.g., Kodachrome, Agfachrome, and Fugichrome. In contrast, color print film names end with the word "color," e.g., Kodacolor and Agfacolor. Each type of color film has a somewhat different color emphasis, which is why certain daylight films, such as Kodachromes, emphasize warm colors such as browns and reds, while Ektachromes emphasize blue tones.

For the best colors, choose a film with a low ASA number. ASA is a measure of film speed (light sensitivity). While faster film speeds permit photography in darker situations, they do so only by compromising color and sharpness.

If your slides tend to be overexposed, try shooting at a lower f stop. This technique, called saturating the film, helps to avoid washed-out overexposures and results in richer colors and more detail. To saturate film in a camera with a manual light meter, simply select a lower f number (or faster shutter speed). Automatic cameras are easily "fooled" into underexposing by a full stop by doubling the ASA setting. Experiment with the saturation technique by alternately shooting the same scene at normal and various saturated exposures.

Before sending your film to a processor, consider that the quality of processing varies from one firm to the next and that bargain processing offers are likely to produce inferior results. If you've invested in quality equipment and taken great care in the field to take the best photos possible, it is inconsistent to send film off for anything less than the best developing and printing processes.

While it might seem curious to consider taking black-and-white pictures of colorful birds, persistent enthusiasm exists among those who regularly expose black-and-white film. The appeal lies in the flexibility of the medium and the creative process, which can range from exposure of the film to developing and printing enlargements at home.

IDENTIFYING PHOTOGRAPHIC PROBLEMS

When photographs return from the processor, carefully sort them into two piles—one for those to keep, the other for rejects. It's as important to study mistakes as it is to admire photographic successes. Here are a few examples of common bird photography problems and some suggestions for increasing your number of "keepers."

103. To avoid taking too many "habitat" photos, examine what you see through the camera's viewfinder and consider the image size before taking the picture. The picture this woman is taking of a heron will have too much extraneous habitat in it.

SPECtacular Photos

Even if birds seem unusually close, they are almost certainly far beyond the reach of cameras without telephoto lenses. Even the longest telephotos provide no guarantee that birds will fill enough of the field of view. While it is not always necessary to fill the field with the subject, too much extraneous "habitat" makes a confusing and boring picture. Learn the limits of your equipment, picturing in your mind what photos will look like when they return from the processor. Before tripping the shutter, take a moment to carefully examine the viewfinder of your single-lens-reflex camera. It will show the exact contents of the photograph and provide an ideal opportunity to recognize spectacular pictures before exposure.

104. This photo of a red-winged blackbird nest has a confusing background and was taken from over the nest. Whenever possible, photograph wildlife closeups by crouching down to eye level with the subject.

105. This excellent photo of a hermit thrush nest was taken at eye level.

Angle of View

The redwing blackbird nest photo (illustration 104) has several conspicuous problems. In addition to not filling enough of the field and having a confusing, busy background, the photographer took the picture while towering over the nest. In contrast, the hermit thrush nest photo (illustration 105) fills the field with the nest and young. Angle of view, however, is the feature that really separates the two photos. To take the thrush photo, the photographer waited in a blind while seated on a camp stool with his tripod-mounted lens focused on the nest. When one of the parent thrushes approached with food, the young lifted their heads, and the photographer was ready.

106. Too much concentration on the subject may lead to a neglect of background detail. The hands holding the bird in this photo compete with the subject for attention.

107. Overconcentration on a subject can also lead to slanting horizons. Be sure to hold your camera level.

Distracting Backgrounds

While looking through the camera viewfinder, examine the background to see that it doesn't distract from or compete with the subject. The photo of the hand-held bird (illustration 106) could have been improved either by filling the field with the bird or by having the holder lift the bird to eye level for human-interest effect. As it is, the background competes with the subject. It is also possible to eliminate some distracting backgrounds by opening the lens diaphragm to reduce the depth of field or by turning the camera to a vertical position.

Slanting horizons (see illustration 107) result from too much concentration on the subject without adequate attention to background detail. While this error is most frequently experienced by photographers shooting from rocking boats, slanting horizons can result whenever photographers fail to hold their cameras level.

108. *This blurred picture of a grackle resulted from shooting at a shutter speed that was too slow.*

109. *Blurred photos result from either slow shutter speeds or inadequate focusing. Fast shutter speeds can "stop" a bird in flight, as in this photo of a flying gull.*

"Soft" Images

There are two usual reasons for blurred or "soft" images. Either the subject moves out of the shallow depth-of-field-focus, or the photo is taken at a shutter speed too slow to stop movement of the lens or bird. To identify pictures blurred from a focus problem look for something either in front of or behind the bird that is in focus. Note that everything in the grackle photo (illustration 108) is out of focus. This clearly results from shooting at a shutter speed too slow to stop the movement of the photographer's telephoto lens.

To "stop" flying birds, set your shutter speed to at least ⅟₅₀₀ second. If possible, attempt to follow the birds and focus in flight. If birds are flying toward you, preset the focus at a known distance and wait for a bird to fly into the prefocused zone. However you do it, prepare to exhaust an abundant supply of film.

110. *The top photo of a swan shows how white birds are easily overexposed. To give the swan the proper exposure, underexpose by a full f stop (as in the bottom photo).*

Overexposures and Underexposures

Extremely light or dark subjects can fool your camera's light meter and result in disappointing exposures. White-plumaged birds such as the swan in illustration 110 reflect more light than darker backgrounds in the same field of view. If your light meter reads the darker tones, the result will be washed-out subjects with poor color and lack of detail. Similarly, dark birds do not reflect as much light as the surrounding field, which frequently leads to underexposures. When possible, avoid these problems by first taking your exposure reading from a "neutral" surface, such as a medium gray or brown color near the subject. To properly expose white birds such as swans, gulls, and egrets, stop your diaphragm down by approximately one full f stop. Likewise, open your lens diaphragm to properly expose dark birds such as crows and blackbirds.

111. The same pileated woodpecker nest was photographed in the morning (top) *and in late afternoon* (bottom). *The photos show how direct lighting changes during the course of the day and can improve your photos.*

Lighting Problems

Although backlighting can sometimes produce dramatic photographic effects, it is usually best to light subjects with direct lighting. Direct lighting highlights details and colors of plumage, and at close distances it lends a sparkle to the bird's eyes. However, it is not always simple to arrange for sunlight to fall conveniently over your shoulder onto the subject.

Flash equipment may provide adequate direct light on overcast days or in situations where a bird seldom leaves the haunts of dark shadows. The flash units must be set within several feet of the subject or they will usually not cast enough light to be of value. If the subject is too distant to use a flash and the sun is out, you can sometimes use a portable mirror to cast light into dark areas, such as under house eaves, to light a nest.

The most satisfying solution is simply to wait for direct light to fall where it is needed. During the day the sun will cover a 180-degree arc as it climbs from the morning eastern sky to the evening western sky. On sunny days direct light will fall on most subjects at one time or another.

Also consider the color of the light. When possible, avoid exposing color film during the heat of the day when the sun is directly overhead (approximately from 11 A.M. to 2 P.M.). Such light is likely to produce harsh contrasts with disturbing glare from light-colored surfaces. Morning or late afternoon light contains warmer tones and will contribute more vital colors.

BIRD BLINDS

Most birds quickly forget observers who hide within the cover of a bird blind. Nearly anything that hides the human form can function as a blind. Watch birds from a car or canoe, on horseback, or from under a loose-fitting poncho and you will find that birds will approach much closer than they would if you did not have such cover. Photographers can use this fact to great advantage by waiting in a blind at places that are likely to attract birds. Bird feeders, birdbaths (particularly those with dripping water), favored feeding areas, and roosts and nests are the most likely places to position a blind.

A window of your home might be the best bird blind for photographing backyard feeder birds (see illustration 112). Ideally, extend the lens through an open window, but if this is not possible, focus through the glass (plate glass gives the least distortion). Bold feeder birds such as chickadees and titmice will readily approach feeders with the photographer standing in full view at the window, but more timid birds may not approach unless the photographer hides from view behind a sheet with only the lens protruding.

Note the direction of the available light and background to determine the best location for a portable blind. Ideally, sunlight should fall directly onto the location where you expect to focus, and the background should not contain distractions, such as houses, telephone lines, or even "busy" vegetation. If you position the blind in a location that is likely to receive strong wind, guy each corner of the blind to the ground with strong cord attached to tent stakes.

A blind positioned near the nest of common backyard birds will provide intimate views and excellent photographic opportunities, but photographers must take special care when photographing active bird nests. To reduce the chance of nest desertion or predation, it is best to wait until the young hatch before posi-

112. Place a bird feeder near a window, which can serve as a bird blind. It won't take long before the birds lose their shyness about your telephoto lens and approach to within easy photographic distance.

tioning the blind. First, set the blind approximately 20 feet from the nest for several days to give the birds a chance to accept this new feature to their landscape. Then, pick up the blind and set it closer, taking care to position it at eye level to the nest. Make certain that your blind can be assembled away from the nest and carried in one piece.

To permit a clear view of the young, it may be necessary to part the vegetation in front of the nest. Rather than cut a clear path and later expose nestlings to weather and predators, use string or masking tape to hold branches and leaves out of the way. Be certain to return carefully all cover to its original position over the nest at the close of each photographic session. When photographing particularly wary species, some photographers find that birds more readily accept the presence of a blind if two people enter the blind. After a few moments one person should leave to give the birds the impression that the blind is vacant.

113. Common birds such as mourning doves (top) *and chipping sparrows* (bottom) *make excellent subjects for bird photographers.* Photos by Michael J. Hopiak

114. *Ice-fishing huts can serve as ideal portable bird blinds to set up in front of nests. To provide adequate lighting for a thicket-nesting bird, this photographer positioned two flash units on the nest.* Photo by Michael J. Hopiak

115. *This photographer perched his blind atop a homemade scaffold to photograph tree-nesting birds at eye level.* Photo by Michael J. Hopiak

116. A bird blind with a convenience perch.

Birds usually have predictable ways of approaching nests and feeders. To obtain bird photos with natural settings, watch for approach routes and prefocus on a branch or other favorite perch. If birds are not perching in a suitable place for photos, position a convenience perch, such as a bare branch, near the nest or feeder. Birds usually adopt such perches as if they had always been a part of the landscape.

117. Watch for the approach routes of birds and prefocus on a branch or other favorite perch, as was done in this excellent photo of a tree sparrow.

Designs for Building a Portable Bird Blind

Portable bird blinds should be no larger than necessary to provide shelter for a seated photographer and a minimum of equipment. The frame should have a minimum of parts and be quick to assemble. Blinds that can be picked up and moved without disassembling permit the photographer to set a blind near a nest with a minimum of disturbance.

Burlap is an ideal material for covering the blind. Its wide weave keeps the blind from overheating and permits the photographer to observe birds approaching from any direction. Some birds, however, can detect movement through the burlap, particularly if the blind is backlit. Canvas and camouflage nylon cloth are also popular blind covers. Regardless of selected material, the blind should have a snug-fitting cover that doesn't blow in the wind and frighten the birds.

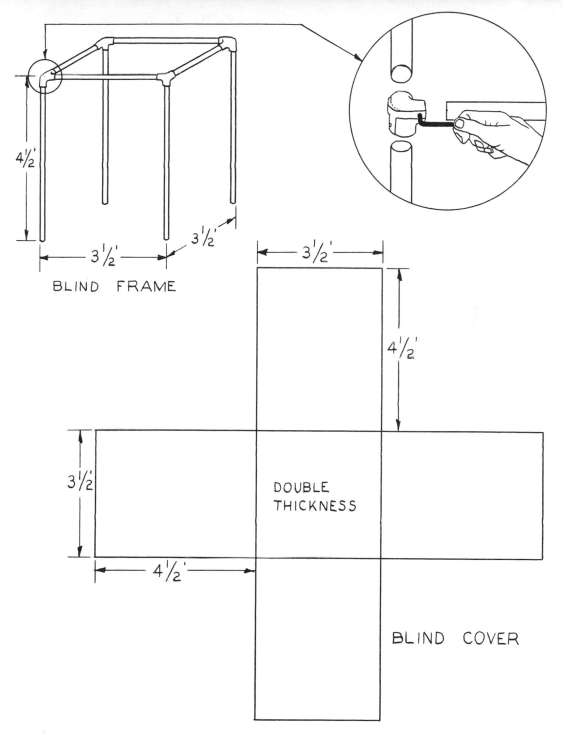

BLIND FRAME

DOUBLE THICKNESS

BLIND COVER

118. Construct this rugged aluminum cube blind from 3/4-inch aluminum conduit, and connect the conduit with Speed-Rail Slip-on fittings. Use three-side outlet elbow connectors to assemble the frame. The only tool necessary to assemble the blind is a 1/8-inch hex wrench. Purchase the conduit and connectors from your local lumber or hardware store. If you cannot find a local source for the connectors, contact The Hollaender Manufacturing Company, 10285 Wayne Avenue, Cincinnati, Ohio 45223.

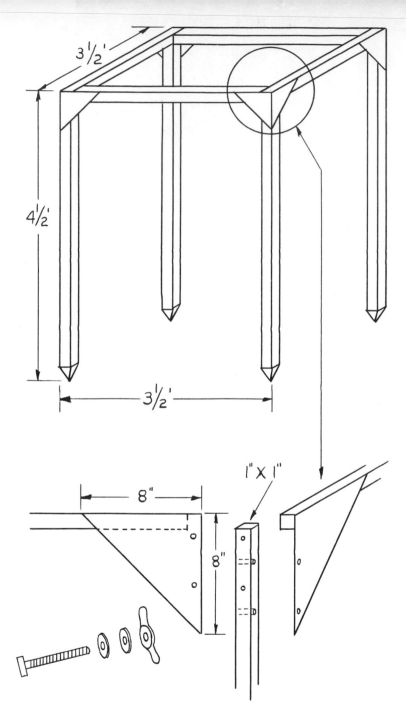

3½'

4½'

3½'

8"

8"

1" X 1"

119. Construct this wooden-frame blind from eight sections of 1-inch-square lumber. Permanently fix triangle braces to each end of the top frame sections, and join the corners by placing 1 3/4-inch stove bolts through the braces and uprights.

BIRD SANCTUARY

NO HUNTING

120. Obtain permission before setting up blinds or stalking birds on private land or wildlife sanctuaries.

BIRD PHOTOGRAPHER'S ETIQUETTE

1. Do not keep parent birds away from their nests and young.
2. Always leave eggs and young in nests.
3. Always leave bird nests exactly as you found them.
4. Cover your trail to bird nests as best you can.
5. Do not lure rare or locally uncommon birds within photographic range using tape recordings.
6. To reduce disturbance to nesting birds, stay in your bird blind for at least one-hour stints.
7. Obtain permission before setting up blinds or stalking birds on private land or wildlife sanctuaries.
8. A camera offers no privileges to push in front of others or to stalk a bird before everyone in a group is finished watching. For these reasons it is usually best to leave your cameras and telephoto lenses at home when joining organized field trips.

BIRD PHOTOGRAPHY PROGRAMS

The Cornell Laboratory of Ornithology has prepared the Home Study Course in Bird Photography, a six-part course for the serious beginner or advanced photographer interested in improving his or her bird photography technique. The topics covered include cinematography (moviemaking), nest photography, high-speed flash technique, and equipment. For more information, write: Cornell Laboratory of Ornithology, 159 Sapsucker Woods Road, Ithaca, N.Y. 14850.

The Photographic Society of America is an international organization with chapters in most major U.S. and Canadian cities. Members meet at regular chapter and regional meetings as well as at an annual convention, during which members attend workshops on photographic techniques and compete for prizes in photographic salons. For further details, write: Photographic Society of America, 2005 Walnut Street, Philadelphia, Pa. 19103.

A FILING SYSTEM FOR BIRD SLIDES

When slides return from the processor, critically examine each slide by spreading them on a light table (see illustration 121). Pull your best slides from the group, and carefully examine the remainder to identify mistakes. If in doubt about sharpness, examine the slides with a 10x hand lens. Then, before the rejects revive too many fond memories, toss them into the wastebasket or store them on a letter spike. The rationale "I paid for it—why not keep it?" will lead to boring slide shows and a hopeless backlog of uncataloged slides. The first step in setting up a slide filing system is to discard mercilessly as many slides as possible.

Aside from photographs of rare species or slides documenting unusual sightings or behaviors, the only slides worth keeping are those you would proudly show friends. If every "record shot" is kept, it soon becomes increasingly difficult to find your best slides, and if slides aren't easily found when you want them, what's the point of taking pictures?

Selection of an appropriate filing system depends on the number and variety of slides in your collection and the amount of time you can apply to cataloging. Keep your system simple so that it doesn't become a chore to catalog and file the slides. Start by discarding rejects, and make it a high priority to file your best slides as soon as they return from the processor.

If you are already behind in filing slides or have yet to start, this section provides a straightforward system that makes slides easy to retrieve and also teaches the organization of bird orders and families.

The estimated 8,600 bird species of the world are organized into 34 orders and 176 families (exact numbers vary according to different authorities). The order and family names serve as an excellent structure for coding slides. With this in mind, assign each order a distinct two-letter code and use this as the prefix in your coding system.

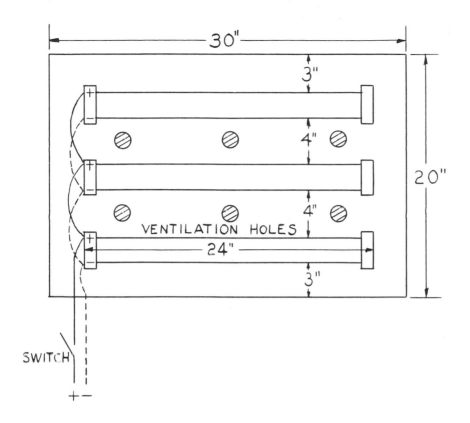

30"

3"

4"

20"

VENTILATION HOLES

4"

24"

3"

SWITCH

+ —

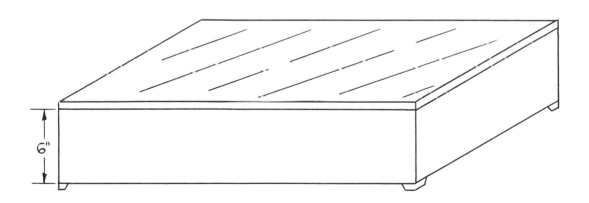

6"

121. A light table is essential for evaluating slides and organizing slide presentations. To construct your own light table, build a box to the dimensions given in this diagram, paint the interior white or silver, and mount three 2-inch-long fluorescent fixtures in a parallel circuit. Cover the box with a piece of white, opaque Plexiglas to complete the table. Commercially built light tables or "previewing screens" are available at most camera shops.

For example, all photos of loons would receive the prefix GA for the order Ga-
viiformes. If the first species of loon photographed happened to be the common
loon, the code for common loon in your system would be GA-1. All future pictures of common loons receive the same number (GA-1). Five years might pass
before you photograph another species of loon, such as the red-throated loon.
This second loon species now receives the code GA-2, and all future slides of red-
throated loons also become GA-2. To retrieve a particular common loon photo,
simply pull all of the GA-1 slides (they are filed together) and spread them on
your light table.

Place the code in the upper right-hand corner of the slide. Use pencil in case
you want to subdivide the order into families at some future time. Record additional data, such as date and location, on the slide mount, but beware that your
data recording does not become too ambitious.

Orders with many birds may require further division because they contain so
many species. To catalog a photo of a yellow-breasted chat (Parulidae), assign it
PA (for the order Passeriformes), P (for the Parulidae family), and the number 1
for the first entry in the system (PA-P-1). All future yellow-breasted chats would
receive the same code, and other warblers would receive the same code with a different numeral, depending on the chronological sequence in which additional species are added to the collection.

*122. To help keep track of your slides, group species that are in the same order or family
on the same page in a loose-leaf catalog. Arrange the pages either alphabetically or taxo-
nomically, following the usual checklist order. This photo shows the warbler page from such
a catalog.*

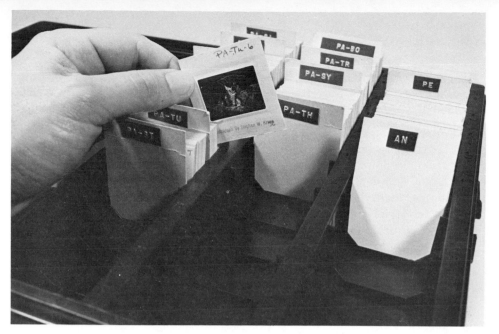

123. Depending on the size of your slide collection, either store the slides in dust-tight slide filing boxes or in large cabinets designed for this purpose. Small filing boxes are available at most camera shops. Avoid the designs that have individual spaces for each slide, since these hold relatively few slides. Cardboard spacers with the various orders and family codes will organize the slides within the boxes. A large cabinet with four drawers, each holding up to thirty-five hundred slides, is available from Steel Fixture Manufacturing Company, 612 East Seventh, P.O. Box 917, Topeka, Kans. 66601.

PRESENTING A SLIDE PROGRAM

If you succeed in taking good bird photos, consider sharing your pictures and experiences with others. One need not be a professional speaker to excite an audience with everyday adventures watching and photographing birds. Almost anyone who makes the effort to photograph birds can step in front of a group and win new recruits to the bird-watching movement.

Bird clubs provide the friendliest and most sympathetic audiences, but you should also attempt to reach people who have not yet discovered the satisfactions of watching birds. Garden clubs, hunting and fishing groups, and camera clubs are likely audiences. Most senior-citizen groups, scout organizations, and school classes will enthusiastically welcome visiting special-interest speakers. Many communities have a speaker's bureau organized through their local chamber of commerce. If your local bird club or Audubon chapter does not already have its own speaker's bureau, consider organizing one from among the bird photographers in your group.

It takes much more than good bird slides to present an enjoyable and informative slide presentation, however, and the following are some techniques to be considered.

1. *Research your subject.* Narratives that rely on such phrases as "this is a" and "here is another" are boring and a waste of everyone's time. Build your presentation around a theme—such as a specific habitat or group of animals. Usually, in-depth, focused approaches to common subjects are best. For example, "Behaviors of Feeder Birds" or "Winter in the Deciduous Forest" are better topics than "My Favorite National Parks" or "The Year I Photographed Six Hundred Bird Species."

2. *Make your presentations entertaining.* Keep them lively, interjecting amusing incidents, but don't strain to be funny. Relevant humor that doesn't degrade people or wildlife is always appropriate. If something funny *did* happen on the way to the lecture, talk about it if it seems appropriate, but don't feel you must start with a funny introduction. Forced humor is worse than silence.

3. *Pick a catchy title* that conjures up positive images, such as a program about nocturnal wildlife entitled "Ghosties and Ghoulies and Things That Go Whomp in the Night," or a lecture on the Neotropics entitled "Flame Birds and Other Tropical Delights." Boring titles suggest boring presentations.

4. *Whenever possible, use your own equipment.* Keep track of the number of hours of life on your projector bulb, replacing the bulbs before they reach the burn-out age. Set up your equipment and prefocus the first slide before the audience enters the room. Arrange to have someone stay with the projector to adjust focus and remove jammed slides. Also arrange for someone to turn off the room lights and bring them up again after your presentation. Use a remote projector advancer so that you can stand in front of the audience rather than behind the projector.

5. *Give a brief introduction* before darkening the room. If there is a lectern, stand in front of it during your introduction so the audience can see what you look like while describing the circumstances behind how you came to take the pictures and make the observations you've come to share. Tell the audience briefly what you will talk about, but don't provide details in the introduction.

6. *Place a 2 × 2-inch cardboard square* (the thickness of a slide) into the slide carrier of a carousel slide projector. This permits you to turn the projector on before starting and then to advance from the "blank" slide to your first slide without asking the projectionist to start the projector. After advancing to the first slide, the blank slide rides in a special slot so that it can drop down into the projector, darkening the screen if you change to a second tray. If you have less than a full tray, put a blank slide into the last slot of the tray to darken the screen so you won't have to ask the projectionist to turn off the projector at the close of the lecture. Never stun your audience by advancing to a glaring, white screen.

7. *Always use a public address system* in an auditorium situation or large room. Check out the system ahead of time to adjust the volume. Remember that sound will not carry as well in a filled auditorium as an empty one, so the microphone volume may have to be adjusted slightly higher for a capacity audience.

8. *Concentrate on speaking distinctly into the microphone.* If the microphone is adjusted properly, you won't have to shout. Look directly at different places in the audience, rather than staring constantly at the screen or your notes. Vary the loudness and pitch of your voice to avoid the monotone effect. Occasionally pause after making an important point to provide emphasis.

9. *Show your enthusiasm for birds* by talking confidently and making every effort to deliver a smooth, lively presentation. Confidence as well as the flow of your "patter" come only with practice. Review the program many times at home, talking out loud to develop a sense of timing and familiarity with the subject.

10. *Rather than explain what a slide illustrates, let the slide illustrate what you say.* For example, picture a bluebird at a nest box. Instead of saying, "You can see that bluebirds nest in homemade boxes," say "Bluebirds nest in homemade boxes." This avoids interpreting the slides and sticks to interpreting the birds. Good slides usually do not need clarification. A good slide narration should stand by itself even without visuals.

11. *Your slides will serve as adequate lecture notes,* if you've practiced adequately. To best use this technique, stand in a position where you can occasionally glance at the screen. If necessary, put prepared notes on large notecards, but never read your presentation.

12. *Most slides should stay on the screen no more than fifteen seconds,* but it is important to vary the projection time from one slide to the next. Presentations for adults using one projector should contain no more than 140 slides and should not last beyond one hour.

13. *Slide presentations for children's groups* should contain no more than seventy slides and should not last more than thirty minutes. Always provide an opportunity for children to ask questions. If you don't know the answers, help them look up the answers in library books. Never talk down to children's groups (or adults). To help maintain order when speaking to a large number of children, keep the room or auditorium lights bright enough so the children can see you.

14. *Whenever possible, show slide sequences that illustrate a particular action pattern*—for example, a heron stalking, catching, and swallowing a fish, or a sequence illustrating care of young at a nest that shows the young at progressive ages. If you have more narration than you can deliver in fifteen seconds for one slide, add additional slides illustrating the same subject viewed from different angles or at different times, but avoid long, repetitious slide series.

15. *Avoid projecting irrelevant slides.* Every slide in the tray should illustrate some point in your narration. Preview your presentation with this in mind and

even eliminate some of your favorite pictures if they do not fit neatly into the theme of the presentation.

16. *Do not acknowledge minor mistakes.* If a slide doesn't drop into place or the projector somehow skips a slide, keep the narration going rather than taking the time to fumble with the projector. Likewise, never back up to a slide that is already past, which will force you to stumble back through the intervening slides to regain your position.

17. *Don't apologize or make excuses for the quality of your photographs.* Comments such as "I didn't have a telephoto lens and I wasn't there to take bird pictures," or "My slides are terrible—I lost the best ones" only focus the attention of the audience on the poor quality and detract from the slides illustrating your narration.

18. *Slide images on the screen needn't be the same size.* To crop out distracting patterns and improve composition, remount problem slides in special mounts with various-sized openings. Another approach is to send slides to a custom color laboratory for cropping so that the image is enlarged to fill the field in a duplicate slide. Such laboratories can produce high-quality copies, and it is safer to show such duplicates than to project your best originals (see following "Resource Materials" section).

19. *To prepare title slides,* position press-on lettering (available in any art supply shop) on a white sheet of paper and photograph the lettering with slide film. Use a close-up lens so that the lettering fills only part of the field (use outdoor light for normal slide film to avoid a yellow glow from indoor lighting). After processing, remove the title from its mount and sandwich it in a new mount over a scenic photo or one with a plain background.

20. *Maps are a must* for slide programs dealing with locations. To avoid walking to the screen and searching for a particular city or other locale, cut out an arrow and lay it in position before you photograph the map with a close-up lens.

21. *End the presentation with a brief review of major points.* The conclusion is the most important part of the program, for it sets the final tone and provides the opportunity to send the audience off with your most important message. If the audience asks for more after your last slide, then you've ended the program at the right time.

Resource Materials

Cardboard mounts with various-sized openings are available from Erie Color Slide Club, Inc., P.O. Box 672, Erie, Pa. 16512.

Recommended custom color labs for duplicating slides are:

Colortek, 330 Newbury Street, Boston, Mass. 02115
Rolph Photography, Inc., 21 Schoen Place, Pittsford, N.Y. 14534
World in Color, P.O. Box 392, Elmira, N.Y. 14902

Duplicate bird slides of most North American birds are available from the Cornell Laboratory of Ornithology, 159 Sapsucker Woods Road, Ithaca, N.Y. 14850.

RECORDING BIRD SOUNDS

Enliven your slide programs by including a few recorded bird sounds. During presentations, keep a tape recorder conveniently positioned to provide an occasional burst of typical sound. While the audience is viewing a particular bird or habitat, such sounds will bring them measurably closer to sharing the feeling of your field experience.

If you succeed in capturing good bird sounds and note field data on your tapes, the recordings might make a valuable contribution to the growing science of animal communication. The songs and calls of many common North American birds are poorly known, and sound libraries welcome recordings to add to their collections. For details about how to contribute tapes and the availability of bird song recordings, see page 118 for a description of the Library of Natural Sounds.

Selecting Equipment

TAPE RECORDERS

To record bird sounds in the field, tape recorders must be battery powered and lightweight. The two basic designs are cassette and reel-to-reel. Most cassette models are lightweight and convenient to use in the field with a price significantly lower than suitable reel-to-reel models. Top-quality cassette recorders cost $900 or more but those priced around $150 can produce excellent recordings to accompany slide presentations, although they may not have the quality necessary to make consistently professional recordings.

The principal drawback to the cassette recorder is the inconvenience of editing field tapes. Unlike the reel-to-reel design where the tape is easily cut and spliced, the tape on a cassette is hidden within the plastic cartridge. To prepare a tape for a program using the cassette system, copy selected sections from field recordings onto a second recorder.

Field recordings made with a reel-to-reel recorder are easier to edit, and selected cuts can be spliced together to produce one tape with a composite of animal sounds. However, prices for reel-to-reel recorders range from approximately $600 for the Uher 4000 Report series to more than $4,000 for certain Nagra recorders. Although reel-to-reel machines are more likely to produce recordings of professional quality with less background noise and a greater frequency range, they are also heavier and bulkier than cassette equipment.

When selecting equipment, be certain your recorder has a digital tape counter. This is essential not only for editing purposes, but for locating tape sections for playbacks in the field. Recorders with a pause switch are ideal for use during slide presentations, because you can let the recorder run and start the sound simply by releasing the pause switch without the loud "clunk" that accompanies most on/off switches.

MICROPHONES

Without a quality microphone, even a high-priced recorder will produce disappointing results; yet it is foolish to purchase a very expensive microphone for an inexpensive recorder. Make certain the microphone matches the tape recorder by checking to see that the impedance rating for the microphone is similar to the manufacturer's specifications for the tape recorder input. (For excellent reviews of various cassette recorders and microphones suitable for recording bird sounds, see articles by T. H. Davis in *Birding*. See Chapter 8.)

There are two basic types of microphones—dynamic and condenser mikes. Dynamic microphones are often preferred for field work because they are rugged and operate without an auxiliary power source. Condenser microphones produce recordings with high-quality sound and a wider frequency range, but their additional circuitry and auxiliary power supply may cause additional maintenance problems. For most bird song recordings, moderate priced microphones of either type will suffice.

When selecting a microphone, consider frequency range and sensitivity. To record most bird sounds, microphones should have a frequency range at least between 50 and 15,000 Hz. To avoid tapes with hissing background sounds, do not use a low-sensitivity microphone unless you use it with a high-quality recorder.

Also consider directional sensitivity when selecting a microphone. Microphones are categorized into one of three groups according to directional sensitivity (see illustration 124). Omnidirectional microphones pick up sound from a 360-degree field. For this reason they are the least valuable for recording bird sounds because they pick up background sounds as readily as the intended bird sounds. Cardioid microphones eliminate most sound from behind the microphone, but they are not nearly as directional as ultradirectional ("shotgun") microphones. Shotgun microphones are ideal for bird sound recording, but the high purchase price of quality models (starting at $300) puts them out of reach for most amateur recorders.

124. Fields of sensitivity for three microphones: (A) omnidirectional; (B) cardioid; (C) ultradirectional ("shotgun"). Redrawn from *Recording Bird Sounds* by J. L. Gulledge

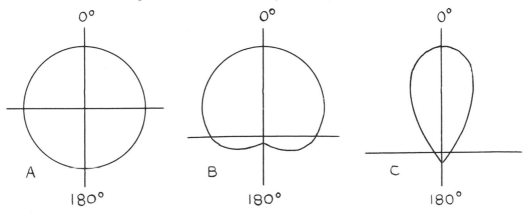

If a bird is vocalizing close to the recorder, even the built-in microphones of certain moderate-priced cassette recorders will succeed in capturing good field recordings. Usually, however, birds are too distant, and there is too much background sound to achieve high-quality recordings with built-in or hand-held microphones other than the shotgun design. Parabolic reflectors are even more effective than shotgun microphones in providing directionality and sensitivity. Although parabolas may introduce some distortion, most serious recorders use them because they effectively reduce background sounds by selectively amplifying the desired bird sounds.

Making Field Recordings

1. To record bird sounds, approach as close as possible without disturbing the bird. It is just as important to be within close range for recording bird sounds as it is for taking quality bird photos.
2. To lure distant territorial birds to close recording range, play back the songs of the same bird or those from a prerecorded bird of the same species. To reduce disturbance, do not use this technique continuously or for rare or locally uncommon birds.
3. Use earphones to aim the parabola with greatest accuracy and to help determine the proper recording level.
4. Bird song is most abundant in early morning, a time when traffic noise and other human disturbances are minimal.
5. Avoid background sounds, such as highway noise and airport traffic, by carefully considering these disturbances when selecting a place to record bird sounds.
6. Recognize that even seemingly insignificant sounds such as a mosquito near the microphone or rain falling on the parabola may totally disrupt the recording.
7. To reduce noises you might make, concentrate on breathing slowly, bracing the parabola, and holding the microphone and earphone cables so that they do not bang against the parabola.
8. To reduce wind noise, avoid pointing the parabola into the wind and place a windscreen over the microphone recording head. Never point a parabola with a mounted microphone toward the sun, as the reflector will concentrate sunlight and may damage the microphone. Likewise, avoid carrying the parabola over your shoulder if the sun is to your back.
9. Leave your camera and telephoto lens behind when making sound recordings. Photography will only compete with and detract from the time and patience necessary to make quality sound recordings.
10. Give your recordings scientific value by including narration on the tape that identifies the bird and provides details about location, date, time, weather, and behavior. Also note the type of recorder and microphone used to make the recording.

A home study
course in
BIRD BIOLOGY

Produced by
The LABORATORY
of ORNITHOLOGY
Cornell University

SEMINARS IN ORNITHOLOGY

Seminar II | THE EXTERNAL BIRD

5

EDUCATIONAL PROGRAMS

Educational programs can provide beginning bird enthusiasts with new and re-warding experiences and promote a greater understanding and appreciation of birds. The educational programs listed in this chapter are divided into three cat-egories: those that welcome participants from all over North America; local or-nithology courses; and international bird tours. Contact the programs that interest you at the addresses given for further information and details.

NORTH AMERICAN PROGRAMS

Correspondence and Museum Programs

Seminars in Ornithology: A Home Study Course in Bird Biology
Laboratory of Ornithology, Cornell University, 159 Sapsucker Woods Road, Ithaca, N.Y. 14850.
    Seminars in Ornithology is a college-level noncredit course in ornithology to be studied at home. It provides a substantial background for the enjoyment and ap-preciation of birds and encourages participants to undertake study projects on

Opposite
125. Sample loose-leaf sheets from Seminars in Ornithology: A Home Study Course in Bird Biology. Photo courtesy of Cornell Laboratory of Ornithology

their own. The course consists of nine seminars, each written by a leading U.S. ornithologist and edited by Olin Sewall Pettingill, Jr., with an emphasis on clarity and readability. A wealth of diagrams, maps, charts, photographs, and drawings by skillful bird artists enrich and enliven the text. The seminars are printed in loose-leaf form, and are organized sequentially, with each of the seminars mailed to the student for study and completion at his or her own pace, before progressing to the next lesson. At the end of each seminar are question sheets that the student completes and returns to the laboratory. An instructor at the laboratory corrects the question sheets, then sends the student the next lesson in the series for completion.

Some of the topics covered are external and internal anatomy, migration, behavior, and bird photography. Participants should have a serious interest in learning more about bird biology. The material presented in the seminars is similar to college courses in introductory ornithology, but the student does not need a background in biology. There is no special entrance requirement, and participants may enroll at any time and progress at their own speed. Write for a free brochure.

The Library of Natural Sounds
Laboratory of Ornithology, Cornell University, 159 Sapsucker Woods Road, Ithaca, N.Y. 14850.

The collection of the Library of Natural Sounds has more than thirty-two thousand recordings of approximately thirty-five hundred species of the world's birds, as well as recordings of many mammals and amphibians. It is the largest and most comprehensive collection of its kind and has become a major biological research resource serving the scientific community, educators, serious amateurs, animal observers, and others who have need of documented recordings of the sounds of vertebrates. Established by Peter Paul Kellogg, the library serves as an archive for general collections that include the sounds made by as many different species of vertebrates as possible and for research collections devoted to studies of particular species or groups of species.

The Library of Natural Sounds welcomes contributing recordists. Special emphasis must be given to recording and documenting the sounds of declining species and those that live in vanishing habitats. The library provides information, advice and technical assistance, and tape to persons willing to record for the collection. Occasionally equipment loans can be arranged. Teams of participants are sometimes organized to explore intensively a selected locality and record the sounds of birds and animals. Participants help subsidize costs and, under the guidance of the team leader, do most of the data gathering and recording. Material from the library is available to anyone who has an interest in animal sounds. Contributing recordists include professional and amateur bird students.

Bird Slide Collection
Photographic Department, Laboratory of Ornithology, Cornell University, 159 Sapsucker Woods Road, Ithaca, N.Y. 14850.

The Bird Slide Collection provides high-quality duplicate slides of all North American birds to the general public and original material to publishers, museums, and universities. Participants assist the program by donating original bird slides. Although the collection is extensive, many species are not well represented. (On request, a list can be provided of the species needed.) Send slides of birds, eggs, or nests with your name, the species, and the date and place the slide was taken (if known) on the slide mount. Also include a letter stating whether you wish to place any restrictions on the use of the slides by the collection. The photographer's name will be printed on the mount of any slide that is sold or duplicated. Photographers receive credit if slides are published. Both professional and amateur bird photographers donate slides to the collection. Slides that are not needed for the collection will be promptly returned.

Birdlab

Friends of the National Zoo, c/o National Zoological Park, Washington, D.C. 20008.

Birdlab provides active learning experiences and reference collections of bird specimens, nests, and eggs. Located in the Bird House at the National Zoo, Birdlab is a resource center of books and objects related to birds. Nests, feathers, and skeletons may be examined there. Learning boxes, observation activities, and worksheets are available as well as a bird information file. Birdlab is staffed by trained volunteers who answer questions and assist the public in their exploration of the world of birds. Birdlab is open Friday, Saturday, and Sunday from twelve to three.

The Naturalist Center

National Museum of Natural History, Mailstop 157, Smithsonian Institution, Washington, D.C. 20560.

The Naturalist Center contains ornithological collections of study skins, skeletons, eggs, and nests. These materials are available for use inside the center by any interested individual (no appointment necessary). The center also houses an extensive ornithological library with identification guides to birds throughout the world. The Naturalist Center contains a wide variety of equipment enabling individual study by visitors in each area of natural history. Microscopes, plant presses, dissecting equipment, hand lenses, and balances are some of the many items available.

The center is designed for use by amateur and professional naturalists. It is intended especially for individual study and is open to anyone over the age of twelve who wants to advance his or her natural history knowledge. The center can be used from Wednesday to Sunday; ask at the information desk at the Natural History Museum.

126. *Birdlab, located in the National Zoo, Washington, D.C., assists the public in their exploration of the world of birds.* Photo courtesy of the National Zoo

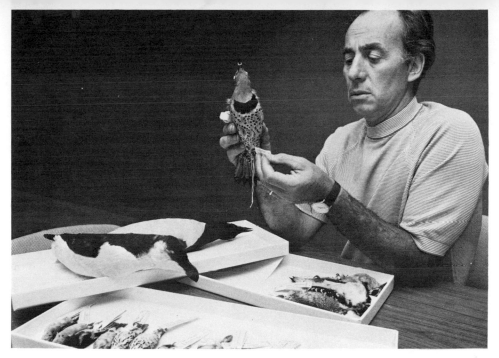

127. *The Naturalist Center, located in the National Museum of Natural History in Washington, D.C., is designed for use by amateur and professional naturalists.* Photo courtesy of the Smithsonian Institution

Field Programs

Audubon Ecology Camps
National Audubon Society, 950 Third Avenue, New York, N.Y. 10022.

Since 1936 Audubon Ecology Camps have offered resident summer courses for adults who want to learn more about the natural environment. Each of the four camps is a completely equipped learning center with dormitories, a dining hall, and modest laboratory facilities. Nearly all classes are held in the outdoors and feature active involvement as the students are immersed in learning about all habitats at each of the four locations. The instructors are enthusiastic naturalists who are well versed in a broad approach to natural science.

Each camp focuses on the natural history of a different ecosystem: Maine—coastal islands with an emphasis on seabirds, marine ecology, and coniferous forests; Connecticut—deciduous forests, ponds, meadows, and seashore; Wisconsin—lakes, wild rivers, prairies, bogs, and forests; and Wyoming—western mountains, coniferous forests, and meadows. The programs interpret birds as part of the total ecosystem. In addition to instruction in bird identification and behavior, the sessions also include instruction in other areas of natural history, such as botany, animal life, geology, and weather.

The courses are open to anyone eighteen or older. Participants represent a diverse background of occupations, in addition to classroom teachers and college students. The only requirement is an enthusiastic interest in learning more about

128. Audubon Ecology Camps offer resident summer courses for adults who want to learn more about the natural environment.

the natural sciences. Interested high-school seniors and college students should investigate the possibility of student assistant positions on the staff. The Maine and Wyoming camps offer four two-week sessions each summer. The Wisconsin camp offers both one- and two-week sessions, and the Connecticut camp offers eight one-week sessions. For an additional fee, university credit is available at all four camps.

Earthwatch
10 Juniper Road, Box 127K, Belmont, Mass. 02178.

Earthwatch is a program that offers opportunities to learn new skills while contributing to serious ornithological and biological research. Participants work with professional ornithologists on such studies as the effects of environmental changes on gull and tern populations near the Great Lakes, population declines

129. *Earthwatch is a program that allows participants to work with professional ornithologists. Participants in this photo are staking out an area for the study of island birds on Nantucket.* Photo by James Baird/Earthwatch

of native Hawaiian birds, and nesting and feeding habits of tropical birds in Panama. Earthwatch cooperates with scientists from the Smithsonian Institution, Point Reyes Bird Observatory, Cornell University Laboratory of Ornithology, and many other research institutions. Teams of participants spend two to three weeks in the field with professional ornithologists, where they are taught all necessary research skills, from mist netting (catching birds in a fine net to band them) and bird banding techniques to identification of particular species.

Earthwatch is open to bird enthusiasts of all ages and occupations with a serious interest in learning scientific techniques. Participants share the costs of the expedition, which generally range from $500 to $800 per person. This contribution and all out-of-pocket expenses are tax deductible.

Atlantic Center Intern Program
Atlantic Center for the Environment, 951 Highland Street, Ipswich, Mass. 01938.

The center provides year-round internships and summer staff positions on bird-related projects. The projects change each year as needs arise, and interns work with the design implementation and evaluation of each project. Projects include conducting bird education programs in rural schools and seabird inventories on remote islands. While orientation and evaluation are conducted at Ipswich, most of the participant's time is spent at a project site in northern Maine or one of the Atlantic provinces. Participants must be at least twenty-one years old, and some experience in ornithology or related natural sciences is preferred. Proficiency in French is helpful, but not required.

Great Gull Island

Helen Hays, Department of Ornithology, American Museum of Natural History, Central Park West at 79th Street, New York, N.Y. 10024.

The American Museum of Natural History conducts long-term studies of common and roseate tern populations on Great Gull Island, located at the eastern end of Long Island Sound. During the spring–summer breeding season, the island is checked daily, nests and eggs are marked, and young terns are banded. Volunteers also trap adult terns on many of the nests. Participants gain experience in recording and processing field data; they must have a serious interest in bird behavior studies and be prepared for routine work, long hours, and primitive living conditions. Inquiries about summer work on Great Gull Island should be submitted by the January or February prior to the breeding season.

Kent Island Ornithological Research Station

Biology Department, Bowdoin College, Brunswick, Maine 04401.

Kent Island is located in the Bay of Fundy, six miles south of Grand Manan Island in the province of New Brunswick, Canada. A variety of research projects are carried out on Kent Island during both the breeding season and fall migration. Depending on available space, visitors are welcome for one week or more and are encouraged to participate in the ongoing research activities, including fall banding. Students, amateurs, and professionals are invited to use the facility to carry out their own research projects. No special qualifications are necessary other than an interest in ornithological research activities. The breeding birds of Kent Island include Leach's storm-petrel, common eider, herring gull, black guillemot, blackpoll warbler, and many others. Habitats include spruce-fir forest, old fields, mud flats, and rocky shore.

Long Point Migration Studies

Dr. David J. T. Hussell, Long Point Bird Observatory, P.O. Box 160, Port Rowan, Ontario N0E 1M0, Canada.

The observatory conducts a year-round program at its headquarters and the shore of Lake Erie and an April–October operation of isolated field stations on Long Point that are manned by volunteers under the direction of a small professional staff. The activities include migration monitoring and a banding program, special studies on birds, educational programs, training workshops, and expenses-paid assistantships for students and other volunteers. Facilities for professional research are also available.

130. Volunteers can participate during the summer in the studies of the common and roseate tern populations on Great Gull Island, at the eastern end of Long Island. Photo by Michael Hale

Manomet Intern Program

Student/Intern Program, Manomet Bird Observatory, Box 936, Manomet, Mass. 02345.

Interns in this program work with staff biologists in ongoing field research programs of the Manomet Bird Observatory for two months or longer. Interns are also encouraged to develop independent research projects or to participate in data analysis of existing projects. Most participants have college-classroom experience, although exceptional high-school students are sometimes accepted. Applicants need not be skillful in bird identification.

Manomet Bander Training Program

Trevor L. Lloyd-Evans, Manomet Bird Observatory, Box 936, Manomet, Mass. 02345.

In this program qualified banders can update their skills with new banding techniques, and volunteers with less banding experience can learn the basics of bird banding under the supervision of professional banders. All participants contribute invaluable help to the long-term data base and current research studies. In addition to a serious interest in bird banding, participants should have some basic ornithology knowledge. Volunteers participate, mainly during spring and autumn migrations, either on a one-day-a-week basis or by staying in limited and rather Spartan accommodations at the observatory for longer periods.

Point Reyes Research Volunteer Program

Dr. David F. DeSante, Point Reyes Bird Observatory, 4990 Shoreline Highway, Stinson Beach, Calif. 94970.

Volunteers in this program usually spend three months at the Palomarin Field Station, located at the Point Reyes National Seashore, thirty miles north of San Francisco. Participants study the population dynamics of selected resident birds

131. Volunteers can learn the basics of bird banding at the Manomet Bander Training Program in Massachusetts. Qualified banders can update their skills with new banding techniques. Photo courtesy of Manomet Bird Observatory

and monitor the coastal migration of land birds, analyzing the effects of local weather patterns on migration. Volunteers participating in the population dynamics research will be assigned a study area and asked to locate all color-banded birds on at least a weekly basis. Participants will also identify the mates of territorial males, map breeding territories, locate nests, and color-band birds. Participants in the migration study assist mist-netting and banding operations as well as the collection of weather data. Participants must have at least moderate experience with field ecology or ornithological research and experience with analysis and write-up of ecological data.

Prince Edward Point Observatory
Mrs. Helen Quilliam, Kingston Field Naturalists, Box 831, Kingston, Ontario K7L 4X6, Canada.

Volunteers at this field station occupying a lighthouse on the shore of Lake Ontario, southwest of Kingston, run banding and study programs, migration censuses, hawk migration counts, and educational and interpretive programs.

ORNITHOLOGY COURSES

The following directory to North American ornithology courses is the result of a survey of more than eight hundred bird clubs and three thousand North American colleges and universities. A very enthusiastic response generated the following listing of over six hundred ornithology courses and clearly demonstrates the widespread popularity and availability of such courses.

The content of credit courses varies greatly, ranging from classroom lectures and laboratory sessions to field courses where the emphasis is on identification and behavior. Write to the listed contacts to determine the content of courses in your community. Most schools offer credit courses to the public through their extramural departments, and it may be possible to audit the course if you are not interested in credit.

Noncredit ornithology courses are frequently sponsored by bird clubs, park districts, and nature centers. The continuing education or adult education departments of many colleges and universities are also frequent sponsors. Noncredit offerings are usually designed specifically for beginning bird-watchers and have an emphasis on field experience. Weekend bird-study courses are becoming increasingly popular. Such programs typically offer a concentrated variety of associated activities such as banding, feeding techniques, painting, photography, and sound recording.

Bird skin collections and taxidermy mounts are useful educational resources that are often neglected by bird-watchers. At such collections you can test your knowledge of familiar birds, examine details such as beaks and feet, and compare similarities in plumage and size that are impossible to determine in the field. Check your local museums and universities to locate the nearest collection.

Organized field courses and field trip programs are an excellent way to meet others who share your interest and at the same time to discover some of the better places to watch birds in your community. Check with your local bird club to find out what field trip programs are available in your community, or contact local recreation departments, park districts, museums, and nature centers.

Alabama

Dr. J. L. Dusi, Department of Zoology and Entomology, Auburn University, Auburn 36830. *Credit course.*

Dr. Paul Yokley, Jr., University of North Alabama, Florence 36830. *Credit course.*

Dr. Charles W. Summerour, Department of Biology, Jacksonville State University, Jacksonville 36265. *Credit course.*

Dr. Elizabeth French, Mobile College, Mobile 36613. *Credit course.*

Department of Biology, University of South Alabama, Mobile 36688. *Credit course.*

Department of Biology, University of Montevallo, Montevallo 35115. *Credit course.*

Dr. Douglas McGinty, Department of Biology, Huntingdon College, Montgomery 36106. *Credit course.*

University of Alabama, Tuscaloosa, Box 1927, University Station 35486. *Credit course.*

Alaska

Biology Department, University of Alaska at Anchorage, 2651 Providence Drive, Anchorage 99504. *Noncredit course.*

Biology Department, Anchorage Community College, 2533 Providence Drive, Anchorage 99504. *Credit course.*

Department of Biological Sciences, University of Alaska, Fairbanks 99701. *Credit course.*

Tanana Valley Community College, University of Alaska, Fairbanks 99701. *Noncredit course.*

Arizona

Department of Biological Sciences, Northern Arizona University, Box 5640, Flagstaff 86011. *Credit course.*

Glen Walsberg, Department of Zoology, Arizona State University, Tempe 85281. *Credit course.*

Arkansas

P. R. Dorris, Henderson State University, H-642, Arkadelphia 71923. *Credit course.*

H. H. Halberg, Pulaski County Audubon Society, 5809 N. Country Club Boulevard, Little Rock 72207. *Noncredit course.*

Fred Burnside, Arkansas Audubon Society, c/o W. O. Grimmett, 24601 Kanis Road, Little Rock 72211. *Noncredit course.*

Aileen L. McWilliam, 711 Magnolia Avenue, Mena 71953. *Noncredit course.*

Robert W. Wiley, Department of Biology, University of Arkansas at Monticello, Monticello 71655. *Credit course.*

Ms. Ellen Neaville, Northwest Arkansas Audubon Society, 821 S. 11th, Rogers 72756. *Noncredit course.*

Dr. H. D. Crawley, Biological Science Department, Arkansas Technical University, Russellville 72801. *Credit course.*

Dr. E. L. Hanebrink, Department of Biological Sciences, Arkansas State University, State University 72467. *Credit course.*

California

Henry E. Childs, Jr., Chaffey College, Alta Loma 91701. *Credit course.*

Biology Department, Pacific Union College, Angwin 94508. *Credit course.*

Golden Gate Audubon Society, 2718 Telegraph Avenue, Suite 206, Berkeley 94705. *Noncredit course.*

Butte College, Pentz Road, Chico 95926. *Noncredit course.*

Department of Biological Sciences, California State University, Chico 95929. *Credit course.*

Mary Wylie, Southwestern College, 900 Otay Lakes Road, Chula Vista 92010. *Noncredit course.*

Dale Keyser, Columbia Junior College, P.O. Box 1849, Columbia 95310. *Noncredit course.*

G. R. Hunsicker, Science Department, Southern California College, 55 Fair Drive, Costa Mesa 92626. *Credit course.*

Ben Hawkins, College of the Redwoods, Eureka 95501. *Credit course.*

Department of Biology, California State University, Fresno 93740. *Credit course.*

Department of Biological Sciences, California State College, Hayward 94542. *Credit course.*

Prof. G. Hunt, Department of Ecology and Biology, University of California at Irvine, Irvine 92717. *Credit course.*

M. D. Kechter, Dean of Sciences, College of Marin, Kentfield 94904. *Credit course.*

R. R. Payne, Department of Biological Science, Biola College, 13800 Biola Avenue, La Mirada 90639. *Credit course.*

Chairman, Biology Department, California State University, Long Beach 90840. *Credit course.*

Los Angeles City College, 855 N. Vermont Avenue, Los Angeles 90029. *Noncredit course.*

Los Angeles Southwest College, 1600 W. Imperial Highway, Los Angeles 90049. *Noncredit course.*

Biology Department, California State University, Los Angeles, 5151 State University Drive, Los Angeles 90032. *Credit course.*

Moore Laboratory of Zoology, Occidental College, Los Angeles 90041. *Credit course.*

Dr. T. R. Howell, Department of Biology, University of California at Los Angeles, Los Angeles 90024. *Credit course.*

Department of Biology and Physical Science, University of California at Los Angeles, 6115 Math-Science Building, Los Angeles 90024. *Noncredit course.*

Foothill College, 12345 El Monte Avenue, Los Altos 94022. *Credit and noncredit courses.*

Dr. J. B. Williams, Department of Natural Science, Pepperdine University, Malibu 90265. *Credit course.*

Prof. Stuart Olson, Department of Biology, Menlo College, Menlo Park 94025. *Credit course.*

Saddleback College, Mission Viejo 92675. *Credit course.*

Life Science Department, Moorpark College, Moorpark 93021. *Credit course.*

Department of Natural Science, Los Angeles Baptist College, Newhall 91322. *Credit course.*

Dr. G. F. Fisler, Biology Department, California State University, Northridge 91330. *Credit course.*

Cliff Nelson, Biology Department, Diablo Valley College, Pleasant Hill 94523. *Credit course.*

Ray Evans, Feather River College, Quincy 95971. *Credit and noncredit courses.*

Carter House Science Museum, P.O. Box 427, Redding 96001. *Noncredit course.*

Shasta College, 1065 N. Old Oregon Trail, Redding 96001. *Credit course.*

Prof. Janet Westbrook, Cerro Casa College, Ridgecrest 93555. *Noncredit course.*

Biology Department, Loma Linda University, La Sierra Campus, Riverside 92515. *Credit course.*

Department of Biology, Sonoma State College, Rohnert Park 94928. *Credit and noncredit courses.*

Biology Department, California State University, Sacramento 95819. *Credit course.*

Keith W. Radford, Department of Biology, San Diego Mesa College, 7250 Mesa College Drive, San Diego 92111. *Credit course.*

Prof. Gerald Collier, Department of Zoology, San Diego State University, San Diego 92182. *Credit course.*

K. M. Hyde, Biology Department, Point Loma College, San Diego 92106. *Credit course.*

Dr. R. Bowman, Department of Biological Sciences, San Francisco State College, San Francisco 94132. *Credit course.*

J. Morlan, Galileo-Marina Community College, 2112 Greenwich Street, San Francisco 94123. *Noncredit course.*

Dr. D. A. Mullen, Department of Biology, University of San Francisco, San Francisco 94117. *Credit course.*

Dan Murphy, 2945 Ulloa Street, San Francisco 94116. *Noncredit course.*

Department of Biological Sciences, San Jose State College, San Jose 95192. *Credit course.*

Ohlone Audubon Society, c/o Mrs. D. Griffin, 2085 150th Avenue, San Leandro 94578. *Noncredit course.*

Santa Barbara Community College Adult Education Center, 914 Santa Barbara Street, Santa Barbara 93101. *Noncredit course.*

Dr. Peter D'Eliseu, Department of Biology, University of Santa Clara, Santa Clara 95053. *Noncredit course.*

Science Department, Allan Hancock College, Santa Maria 93454. *Credit and noncredit courses.*

Jon Winter, Life Science Department, Santa Rosa Junior College, Santa Rosa 95446. *Credit course.*

Department of Biological Sciences, University of the Pacific, Stockton 95211. *Credit course.*

Dr. P. A. Nickel, Department of Biological Sciences, California Lutheran College, Thousand Oaks 91360. *Credit course.*

Marin Audubon Society, P.O. Box 441, Tiburon 94920. *Noncredit course.*

Ken Beatty, Biology Department, College of the Siskiyous, Weed 96094. *Credit course.*

Los Angeles Harbor College, 1111 Figueroa Place, Wilmington 90744. *Noncredit course.*

S. A. Karlin, Department of Life Sciences, Los Angeles Pierce College, Woodland Hills 91364. *Noncredit course.*

Colorado

Dr. Veryl F. Keen, Department of Biology, Adams State College, Alamosa 81102. *Credit course.*

Drs. O. Williams, C. Bock, or A. Cruz, Department EPO Biology, University of Colorado, Boulder 80309. *Credit course.*

James Enderson, Department of Biology, Colorado College, Colorado Springs 80903. *Credit course.*

Interpretive Staff, El Paso County Parks Solar Trails Center, 245 Bear Creek Road, Colorado Springs 80906. *Noncredit course.*

Department of Biology, Regis College, W. 50th and Lowell Boulevard, Denver 80221. *Credit course.*

Biology Department, University of Colorado, Denver Center, Denver 80202. *Credit course.*

Dr. R. Cohen, Department of Biology, Metropolitan State College, 1006 11th Street, Denver 80204. *Credit course.*

Department of Biological Sciences, University of Denver, Denver 80208. *Credit course.*

Office of Continuing Education, University of Denver, Denver 80208. *Noncredit course.*

Division of Science and Mathematics, Community College of Denver, 12600 W. 6th Avenue, Golden 80401. *Credit course.*

Department of Biological Sciences, Mesa College, Grand Junction 81501. *Credit course.*

Science-Math Division, Aims Community College, Box 69, Greeley 80631.
Credit course.

Life Science Department, University of Southern Colorado, Pueblo 81001.
Credit course.

Dr. Gerald Dotson, Community College of Denver, 3645 W. 112th Avenue,
Westminster 80030. *Credit course.*

Connecticut

Dr. M. E. Somers, Biology Department, University of Bridgeport, Bridgeport
06602. *Credit course.*

M. G. Bull, Natural History Services, Connecticut Audubon Society, 314
Unquowa Road, Fairfield 06430. *Noncredit course.*

Dr. Richard Bernard, Quinnipiac College, P.O. Box 84, Hamden 06518.
Credit course.

Dr. N. Proctor, Department of Biology, Southern Connecticut State College,
New Haven 06515. *Credit course.*

Dr. William J. Barry, Department of Biology, Connecticut College, New London
06320. *Credit course.*

Dr. George A. Clark, Jr., Sys. and Evol. Biology, U43, University of Connecticut,
Storrs 06268. *Credit course.*

Delaware

Delaware Nature Education Society, Box 700, Hockessin 19707. *Noncredit course.*

District of Columbia

Dr. V. C. Guerrero, Wildlife Management, College of Life Sciences, University
of District of Columbia, 4200 Connecticut Avenue N.W., 20008. *Credit course.*

Florida

University of Miami, Department of Biology, P.O. Box 249118, Coral Gables
33124. *Credit course.*

Dr. D. Stock, Department of Biology, Stetson University, Deland 32720.
Credit course.

Dr. T. C. Emmel, Department of Zoology, University of Florida, Gainesville
32611. *Credit course.*

SFCC Bio-Parks Program, 3000 N.W. 83rd Street, Gainesville 32602. *Credit course.*

Chairman, Department of Natural Sciences, University of North Florida,
Jacksonville 32216. *Credit course.*

Dr. R. W. Loftin, University of North Florida, Box 17074, Jacksonville 32216.
Credit course.

Dr. W. K. Taylor, Department of Biological Sciences, University of Central
Florida, Box 2500, Orlando 32816. *Credit course.*

Sanibel-Captiva Audubon Society, Sanibel 33957. *Noncredit course.*

Dr. F. C. James, Department of Biological Sciences, Florida State University, Tallahassee 32306. *Credit course.*

Department of Biology, LIF 169, University of South Florida, Tampa 33620. *Credit course.*

Georgia

G. Schmalz, Fernbank Science Center, 156 Heaton Park Drive N.E., Atlanta 30307. *Credit course.*

Oglethorpe University, Continuing Education, 4484 Peachtree Road N.E., Atlanta 30319. *Noncredit course.*

Dr. E. K. Urban, Department of Biology, Augusta College, Augusta 30904. *Credit course.*

Biology Department, Columbus College, Columbus 31907. *Credit course.*

Biology Department, Georgia Southern College, Statesboro 30458. *Credit course.*

Dr. C. Connell, Department of Biology, Valdosta State College, Valdosta 61601. *Credit and noncredit courses.*

Hawaii

Chairman, Department of Zoology, University of Hawaii, Honolulu 96822. *Credit course.*

Idaho

Department of Biology, Boise State College, Boise 83725. *Credit course.*

Dr. R. D. Bratz, College of Idaho, Caldwell 83605. *Credit course.*

Dr. Tom Urquhart, Lewis-Clark State College, Lewiston 83501. *Noncredit course.*

Dr. D. Johnson, Department of Biological Sciences, University of Idaho, Moscow 83843. *Credit course.*

Department of Biology, Northwest Nazarene College, Nampa 83651. *Credit course.*

Department of Biology, Idaho State University, Pocatello 83209. *Credit and noncredit courses.*

Ririe Godfrey, Biology Department, Ricks College, Rexburg 83440. *Credit course.*

Illinois

T. N. Ingram, Eagle Valley Environmentalists, Box 155, Apple River 61001. *Weekend workshop.*

Dr. C. Zimmerman, Biology Department, Aurora College, Aurora 60506. *Credit course.*

Department of Zoology, Eastern Illinois University, Charleston 61920. *Credit course.*

Department of Biology, Northeastern Illinois University, 5500 N. St. Louis Avenue, Chicago 60625. *Credit course.*

Dr. N. H. Jenson, Department of Biology, Millikin University, Decatur 62523. *Credit course.*

Department of Biological Sciences, Northern Illinois University, DeKalb 60115. *Credit course.*

Department of Biological Sciences, Southern Illinois University, Edwardsville 62026. *Credit course.*

Department of Biology, Elmhurst College, Elmhurst 60126. *Credit course.*

Dr. G. T. Girard, Biology Department, Principia College, Elsah 62028. *Credit course.*

Dr. William Zales, Joliet Junior College, 1216 Houbolt Road, Joliet 60436. *Noncredit course.*

Dr. T. R. Anderson, Division of Science, McKendree College, Lebanon 62254. *Credit course.*

Biology Department, Illinois Benedictine College, 5700 College Road, Lisle 60532. *Credit course.*

Dr. E. C. Franks, Department of Biological Sciences, Western Illinois University, Macomb 61455. *Credit course.*

Jan Wiseman, Community Education and Services, Kichwanhee College, Malta 60150. *Noncredit course.*

Drs. Birkenholz and Thompson, Department of Biological Sciences, Illinois State University, Normal 61761. *Credit and noncredit courses.*

Dr. R. G. Bjorklund, Department of Biology, Bradley University, Peoria 61625. *Credit course.*

G. H. Schneider, Department of Biological Sciences, Quincy College, Quincy 62301. *Credit course.*

Norine Mahlburg, 813 N. Main Street, Rockford 61103. *Noncredit course.*

Head, Department of Biology, Augustana College, Rock Island 61201. *Credit course.*

Indiana

Dr. J. D. Goodman, Department of Biology, Anderson College, Anderson 46011. *Credit course.*

Department of Biology, Indiana University, Bloomington 47405. *Credit course.*

Charles Cardinal, Biology Department, Ancilla College, Donaldson 46513. *Credit and noncredit courses.*

Dr. Gary Lee Tieben, Saint Francis College, Fort Wayne 46808. *Credit course.*

Continuing Education, Indiana University and Purdue University, Fort Wayne 46805. *Noncredit course.*

Dr. James Curry, Department of Biology, Franklin College, Franklin 46131. *Credit course.*

Dr. J. Dan Webster, Hanover College, Hanover 47243. *Credit course.*

Chairman, Biology Department, Huntington College, Huntington 46750. *Credit course.*

Zoology Department, Butler University, 4600 Sunset Avenue, Indianapolis 46208. *Credit course.*

Alfred (Bud) Starling, 7207 Lafayette Road, Indianapolis 46278. *Noncredit course.*

Education Department, Indianapolis Museum of Art, 1200 W. 38th Street, Indianapolis 46208. *Noncredit course.*

Mr. Mike Goff, Biology Department, Marion College, Marion 46952. *Credit course.*

Department of Biology, Ball State University, Muncie 47306. *Credit course.*

Dr. D. A. Jones, Department of Biology, Saint Joseph's College, Rensselaer 47978. *Noncredit course.*

William H. Buskirk, Biology Department, Earlham College, Richmond 47374. *Credit course.*

Prof. Mary Ann Morse, Indiana University East, Richmond 47374. *Credit course.*

Drs. Proffitt and Tamar, Department of Life Sciences, Indiana State University, Terre Haute 47809. *Credit course.*

Prof. C. H. Krexeler, Department of Biology, Valparaiso University, Valparaiso 46383. *Credit course.*

F. H. Montague, Jr., Department of Forestry, Purdue University, West Lafayette 47907. *Credit course.*

Dr. J. C. Jankowski, Calumet College, 2400 New York Avenue, Whiting 46394 *Credit course.*

Iowa

Chairman, Animal Ecology Department, Iowa State University, Ames 50011. *Credit course.*

Scott Community College (Putnam Museum), Belmont Road, Bettendorf 52722. *Noncredit course.*

R. A. Young, Department of Math/Sciences, Kirkwood Community College, P.O. Box 2068, Cedar Rapids 52406. *Credit course.*

Office of Admissions, Coe College, Cedar Rapids 52402. *Credit course.*

Ms. Dot Hinman, Continuing Education, Coe College, Cedar Rapids 52402. *Noncredit course.*

Department of Biology, Simpson College, Indianola 50125. *Credit course.*

Iowa Valley Community College District, 22 W. Main, Marshalltown 50158. *Noncredit course.*

Dr. David Lyon, Cornell College, Mount Vernon 52314. *Credit course.*

Dr. G. Eiben, Associate Professor of Biology, Wartburg College, Waverly 50677. *Credit course.*

Kansas

Biology Department, Baker University, Baldwin 66006. *Credit course.*

Department of Biology, Saint Mary of the Plains College, Dodge City 67801. *Credit course.*

Division of Biological Sciences, Emporia Kansas State College, Emporia 66801. *Credit course.*

Prof. Sherm C. Nystrom, Barton College, Great Bend 67530. *Noncredit course.*

C. A. Ely, Department of Biology, Fort Hays Kansas State College, Hays 67601. *Credit course.*

J. C. Johnson, Jr., Biology Department, Pittsburg State University, Pittsburg 66762. *Credit course.*

Dr. L. Bussjaeger, Department of Biology, Kansas Newman College, 3100 McCormick, Wichita 67213. *Credit course.*

Department of Biology, Southwestern College, Winfield 67156. *Credit course.*

Kentucky

Registrar, Berea College, Berea 40404. *Credit course.*

H. E. Shadowen, Department of Biology, Western Kentucky University, Bowling Green 42101. *Credit course.*

Dr. T. C. Rambo, Department of Biological Sciences, Northern Kentucky University, Highland Heights 41076. *Credit course.*

Dr. R. W. Barbour, School of Biology, University of Kentucky, Lexington 40506. *Credit course.*

Department of Biology, University of Louisville, Louisville 40208. *Credit course.*

Prof. F. Busrow, Department of Biology, UPO Box 1352, Morehead State University, Morehead 40351. *Credit course.*

Dr. R. Goetz, Department of Biology, Murray State University, Murray 42071. *Credit course.*

Dr. Cecile Boehmer, Alice Lloyd College, Pippa Passes 41844. *Credit course.*

Prof. A. L. Whitt, Department of Biological Sciences, Eastern Kentucky University, Richmond 40475. *Credit and noncredit courses.*

Louisiana

Gayle Strickland, 5116 Highland Road, Baton Rouge 70808. *Noncredit course.*

Mr. Harland D. Guillory, Louisiana State University at Eunice, P.O. Box 1129, Eunice 70535. *Credit course.*

Dr. M. B. Eyster, Biology Department, University of Southwestern Louisiana, P.O. Box 42451, Lafayette 70504. *Credit course.*

Dr. D. T. Kee, Department of Biology, Northeast Louisiana University, Monroe 71209. *Credit course.*

Department of Biological Sciences, Northwestern State University of Louisiana, Natchitoches 71457. *Credit and noncredit courses.*

Department of Biological Sciences, University of New Orleans, New Orleans 70122. *Credit course.*

Dr. J. Goertz, Zoology Department, Box 5797, Louisiana Tech University, Ruston 71272. *Credit and noncredit courses.*

Maine

Androscoggin Historical Society, Court Street Building, County Building, Auburn 04210. *Noncredit course.*

Will Russell and William Drury, College of the Atlantic, Bar Harbor 04609. *Credit course.*

Prof. C. E. Huntington, Department of Biology, Bowdoin College, Brunswick 04011. *Credit course.*

Dr. R. L. Martin, Department of Biology, University of Maine, Farmington 04938. *Credit course.*

Prof. R. Chute, Biology Department, Bates College, Lewiston 04240. *Study skin collection.*

Department of Zoology, University of Maine, Orono 04473. *Credit and noncredit courses.*

Chairman of Environmental Science, Unity College, Unity 04988. *Credit course.*

Maryland

Biology Department, Loyola College, 4501 N. Charles Street, Baltimore 21210. *Credit and noncredit courses.*

S. W. Simon, Natural Science, Catonsville Community College, 800 S. Rolling Road, Catonsville 21228. *Credit course.*

Department of Biology, Hood College, Frederick 21701. *Credit course.*

K. Y. Hodgdon, Allegany Community College, 945 Weires Avenue, La Vale 21502. *Credit course.*

M. D. Ferrier, Institute of Biology, Garrett Community College, McHenry 21541. *Noncredit course.*

Dr. E. J. Willoughby, Division of Natural Science, St. Marys College of Maryland, St. Marys City 20686. *Credit course.*

Dr. P. D. Creighton, Department of Biological Sciences, Towson State University, Towson 21204. *Credit course.*

Massachusetts

Athol Bird and Nature Club, 494 S. School Street, Athol 01331. *Study skin collection and noncredit course.*

Dr. S. Duncan, Biology Department, Boston University, 2 Cummington Street, Boston 02215. *Credit course.*

Dr. J. Hatch, Biology Department, University of Massachusetts, Boston 02125. *Credit course.*

Biology Department, Boston State College, 625 Huntington Avenue, Boston 02115. *Credit course.*

Division of Continuing Education, Framingham State College, Framingham 01701. *Credit course.*

Education Department, Manomet Bird Observatory, Box 936, Manomet 02345. *Noncredit course.*

Dr. J. C. Kricher, Biology Department, Wheaton College, Norton 02703. *Credit course.*

Dr. S. Smith, Department of Biological Sciences, Mount Holyoke College, South Hadley 01075. *Credit course.*

Dr. G. Wasmer, Department of Biology, Atlantic Union College, South Lancaster 01561. *Credit course.*

Biology Department, Westfield State College, Westfield 01085. *Credit course.*

Winter Study Program, Williams College, Williamstown 01267. *Credit course.*

Michigan

Prof. C. Weatherby, Department of Biology, Adrian College, Adrian 49221. *Credit and noncredit courses.*

Biology Department, Albion College, Albion 49224. *Credit course.*

Whitehouse Nature Center, Albion College, Albion 49224. *Noncredit course.*

Ronald Ward, Department of Biology, Grand Valley State Colleges, Allendale 49401. *Credit course.*

Dr. Lester Eyer, Alma College, Alma 48801. *Credit course.*

T. D. Howes, Genessee Audubon Club, 8424 Perry Road, Atlas 48411. *Noncredit course.*

D. Wujek or K. Wittenstrom, Oakland Community College, Auburn Heights 48057. *Credit course.*

Mr. E. Barnett, Institute of Biological Science, Kellogg Community College, Battle Creek 49016. *Credit course.*

Dr. A. C. Thoresen, Biology Department, Andrews University, Berrien Springs 49104. *Credit course.*

Department of Biology, Ferris State College, Big Rapids 49307. *Credit course.*

Admissions, Oakland Community College, 2480 Updyke, Bloomfield Hills 48013. *Credit course.*

Cranbrook Institute of Science, 500 Lone Pine Road, Bloomfield Hills 48013. *Study skin collection and noncredit course.*

O. G. Gelderloos, University of Michigan, 4901 Evergreen Road, Dearborn 48128. *Credit course.*

Dr. W. Thompson, Department of Biological Sciences, Wayne State University, Detroit 48201. *Credit course.*

Dr. D. Beaver, Department of Zoology, Michigan State University, East Lansing 48824. *Credit course.*

Biology Department, Oakland Community College, Orchard Ridge Campus, 27055 Orchard Lake Road, Farmington 48018. *Noncredit course.*

Ray Gates, Grand Rapids Baptist College, Grand Rapids 49505. *Credit course.*

Dr. J. Catenhusen, Biology Department, Hillsdale College, Hillsdale 49242. *Credit course.*

Department of Biology, Hope College, Holland 49423. *Credit and noncredit courses.*

Dr. R. Brewer, Biology Department, Western Michigan University, Kalamazoo 49008. *Credit course.*

Registrar, Kalamazoo College, Kalamazoo 49007. *Credit course.*

Schoolcraft College Office of Continuing Education, 18600 Haggerty Road, Livonia 48152. *Noncredit course.*

Dr. William Robinson, Department of Biology, Northern Michigan University, Marquette 49855. *Credit course.*

Dr. M. Hamas, Biology Department, Central Michigan University, Mount Pleasant 48859. *Credit course.*

Dr. R. C. Fleming, Department of Biology, Olivet College, Olivet 49076. *Credit course.*

Dr. R. Reilly, Bio-Chem Department, Lake Superior State College, Sault Sainte Marie 49783. *Credit course.*

Dr. E. E. Whiteman, Spring Arbor College, Spring Arbor 49283. *Credit course.*

Sci-Math Department, Oakland Community College, Highlands Lakes Campus, 7350 Cooley Lake Road, Union Lake 48085. *Credit course.*

Dr. H. H. Caswell, Jr., Biology Department, Eastern Michigan University, Ypsilanti 48197. *Credit course.*

Minnesota

Dr. William Brooks, ACM Summer Wilderness Field Station, Ely 55731. *Credit course.*

Leland H. Grim, Rainy River Community College, International Falls 56649. *Credit and noncredit courses.*

Records Office, Inver Hills Community College, 8445 College Trail, Inver Grove 55075. *Noncredit course.*

Dr. M. J. Fryendall, Department of Biology, Mankato State University, Mankato 56001. *Credit course.*

Department of Biology/Earth Space Science, Southwest State University, Marshall 56258. *Credit course.*

Prof. H. Tordoff, 301 Bell Museum of Natural History, 10 Church Street, S.E., University of Minnesota, Minneapolis 55455. *Credit course.*

Biology Department, Moorhead State University, Moorhead 56560. *Credit course.*

Department of Biology, Concordia College, Moorhead 56560. *Credit course.*

Dr. A. Petersen, Biology Department, St. Olaf College, Northfield 55057. *Credit course.*

Wood Lake Nature Center, 735 W. Lake Shore Drive, Richfield 55423. *Noncredit course.*

Dr. A. Grewe, Department of Biological Science, St. Cloud State University, St. Cloud 56301. *Credit course.*

Biology Department, Concordia College, St. Paul 55104. *Credit course.*

The Science Museum of Minnesota, 505 Wabasha Street, St. Paul 55102. *Noncredit course.*

Dr. C. R. Samuelson, Northland Community College, Highway One East, Thief River Falls 56701. *Credit course.*

Dr. J. Opsahl, Department of Biology, Winona State University, Winona 55987. *Credit course.*

Dr. R. A. Faber, Biology Department, Saint Mary's College, Winona 55987. *Credit course.*

Mississippi

Chairman, Department of Biological Sciences, Delta State University, Cleveland 38733. *Credit course.*

Chairman, Biology Department, Mississippi College, Clinton 39058. *Credit course.*

Dr. J. A. Jackson, Department of Biological Sciences, Box 6Y, Mississippi State University, State College 39762. *Credit and noncredit courses.*

Department of Biology, University of Mississippi, University 38677. *Credit course.*

Missouri

Charles Laun, Stephens College, Columbia 65201. *Credit and noncredit courses.*

Biology Department, Central Methodist College, Fayette 65248. *Credit course.*

Missouri Department of Conservation, P.O. Box 180, Jefferson City 65102. *Noncredit course.*

Peter Goldman, Biology Department, Northeast Missouri State University, Kirksville 63501. *Credit course.*

Dr. D. A. Easteria, Department of Biology, Northwest Missouri State University, Maryville 64468. *Credit and noncredit courses.*

Dr. D. W. Davis, Department of Biology, School of the Ozarks, Point Lookout 65726. *Credit course.*

Biology Department, Harris-Stowe State College, St. Louis 63103. *Credit course.*

Dr. N. J. Sappington, Missouri Baptist College, 12542 Conway Road, St. Louis 63141. *Credit course.*

Jack Williams, State Fair Community College, Sedalia 65301. *Credit course.*

Dr. T. Stombaugh, Department of Life Sciences, Southwest Missouri State University, Springfield 65804. *Credit course.*

Admissions, Central Missouri State University, Warrensburg 64093. *Credit course.*

Montana

Dr. N. Schoenthal, Department of Biology, Eastern Montana College, Billings 59101. *Credit course.*

Dr. R. L. Eng, Biology Department, Montana State University, Bozeman 59717. *Credit course.*

Dr. Dan Block, Western Montana College, Dillon 59725. *Credit course.*

Department of Zoology, University of Montana, Missoula 59812. *Credit course.*

Center for Continuing Education, University of Montana, Missoula 59812. *Noncredit course.*

Nebraska

Dr. H. R. Lawson, Biology Department, Chadron State College, Chadron 69337. *Credit course.*

Dr. R. Quinn, Nebraska Wesleyan University, 50th and St. Paul Streets, Lincoln 68504. *Noncredit course.*

School of Life Sciences, 348 MHLS, University of Nebraska—Lincoln, Lincoln 68588. *Credit course.*

Nevada

Chairman, Department of Biological Sciences, University of Nevada—Las Vegas, Las Vegas 89154. *Credit course.*

Coordinator, Division of Continuing Education, University of Nevada—Las Vegas, Las Vegas 89154. *Noncredit course.*

Dr. F. Ryser, Department of Biology, University of Nevada—Reno, Reno 89557. *Credit course.*

New Hampshire

Dave Carlisle, c/o Berlin Unc-Tech College, Berlin 03570. *Credit course.*

Audubon Society of New Hampshire, 3 Silk Farm Road, Concord 03301. *Credit and noncredit courses.*

Department of Zoology, University of New Hampshire, Durham 03824. *Credit course.*

Squam Lakes Science Center, P.O. Box 146, Holderness 03245. *Noncredit course.*

Dr. Harold Goder, Keene State College, Keene 03431. *Credit course.*

Natural Sciences Department, Plymouth State College, Plymouth 03264. *Credit and noncredit courses.*

New Jersey

Scherman Sanctuary, Box 693, Bernardsville 07924. *Study skin collection and noncredit course.*

Dr. R. Hastings, Department of Biology, Rutgers University, Camden 08102. *Credit course.*

Cape May Weekend, New Jersey Audubon Society, 790 Ewing Avenue, Franklin Lakes 07417. *Weekend workshop.*

Dr. R. Raimist, Life Sciences Department, Glassboro State College, Glassboro 08028. *Credit course.*

Dr. Charles Leck, Department of Zoology, Rutgers University, New Brunswick 08903. *Credit course.*

Dr. C. L. Churchill, Burlington County College, Pemberton Browns Mills 08060. *Credit course.*

Division of Natural Science, Stockton State College, Pomona 08240. *Credit course.*

Department of Education, 205 W. State Street, P.O. Box 1868, Cultural Center, Trenton 08625. *Study skin center and noncredit course.*

Dr. J. Mahoney, Biology Department, Kean College of New Jersey, Union 07083. *Credit course.*

Dr. S. M. Kuhnen, Biology Department, Montclair State College, Upper Montclair 07043. *Credit course.*

Dr. L. E. Spiegel, Biology Department, Monmouth College, West Long Branch 07764. *Credit course.*

New Mexico

Dr. D. Ligon, Department of Biology, University of New Mexico, Albuquerque 87131. *Credit course.*

W. Prentice, Apache Elementary School, 12800 Copper N.E., Albuquerque 87123. *Study skin collection and noncredit course.*

R. J. Raitt, Department of Biology, New Mexico State University, Las Cruces 88003. *Credit and noncredit courses.*

Dr. A. L. Gennaro, Eastern New Mexico University, Portales 88130. *Credit course.*

Department of Biological Sciences, Western New Mexico University, Silver City 88061. *Credit course.*

New York

Department of Biological Sciences, State University of New York at Albany, 1400 Washington Avenue, Albany 12222. *Credit course.*

Beaver Lake Nature Center, 8477 E. Mude Lake Road, Baldwinsville 13207. *Study skin collection and noncredit course.*

Floyd R. West, Broome Technical Community College, Binghamton 13902. *Noncredit course.*

R. E. Andrus and J. Shepherd, Biology Department, State University of New York at Binghamton, Binghamton 13901. *Credit course.*

Dr. Ronald C. Dilcher, Biology Department, State University College at Brockport, Brockport 14420. *Credit course.*

Buffalo Museum of Science, Humboldt Parkway, Buffalo 14211. *Noncredit course.*

Dr. R. C. Stein, Department of Biology, State University College at Buffalo, Buffalo 14222. *Credit course.*

Dr. C. White, Community College of the Finger Lakes, Canandaigua 14424. *Credit course.*

Biology Department, St. Lawrence University, Canton 13617. *Credit course.*

Vanderbilt Museum, Centerport 11721. *Study skin collection.*

Dr. J. Gustafson, Department of Biological Sciences, State University College at Cortland, Cortland 13045. *Credit course.*

Prof. Michael Smiles, Hale Hall, State University College at Farmingdale, Farmingdale 11735. *Noncredit course.*

Prof. Max Hecht, Department of Biology, Queens College, Flushing 11367. *Credit course.*

Dr. James Parker, Biology Department, State University College at Fredonia, Fredonia 14063. *Credit and noncredit courses.*

Department of Biology, State University of New York at Geneseo, Geneseo 14454. *Credit course.*

Leisure Learning School, State University of New York at Geneseo, Geneseo 14454. *Noncredit course.*

Dr. R. A. Ryan, Biology Department, Hobart and William Smith Colleges, Geneva 14456. *Credit course.*

Dr. J. D. Lyon, Adirondack Community College, Glens Falls 12801. *Credit course.*

Dr. Robert Johnson, Hofstra University, Hempstead 11550. *Credit course.*

Director of Continuing Education, Columbia-Greene Community College, Box 1000, Hudson 12534. *Study skin collection and noncredit course.*

Dr. John L. Confer, Biology Department, Ithaca College, Ithaca 14850.
Credit course.

Summer Seminars in Ornithology, Cornell University, 626B Thurston Avenue,
Ithaca 14853. *Noncredit course.*

Registrar, Day Hall, Cornell University, Ithaca 14853. *Credit course.*

Laboratory of Ornithology, Cornell University, 159 Sapsucker Woods Road,
Ithaca 14850. *Noncredit course.*

Shoals Marine Laboratory (Maine summers), G-14 Stimson Hall, Cornell
University, Ithaca 14853. *Credit course.*

Dr. K. Cooper, York College, Jamaica 11451. *Credit course.*

Biology Department, Jamestown Community College, Jamestown 14701.
Credit course.

Dr. James White, Keuka College, Keuka Park 14478. *Credit course.*

Department of Biology, State University of New York at New Paltz, New Paltz
12562. *Credit and noncredit courses.*

Dr. J. Miller, Hartwick College, Oneonta 13820. *Credit course.*

G. R. Maxwell, Department of Zoology, State University of New York at
Oswego, Oswego 13126. *Credit course.*

Biology Department, State University College at Potsdam, Potsdam 13676.
Credit course.

Ralph T. Waterman Bird Club, Inc., c/o Mrs. Alice T. Jones, Knolls Road,
Poughkeepsie 12601. *Noncredit course.*

William Jacobs, Dutchess Community College, Poughkeepsie 12580.
Noncredit course.

Dr. James M. Utter, Division of Natural Sciences, State University of New York
at Purchase, Purchase 10577. *Credit course.*

Dr. S. W. Eaton, Biology Department, Saint Bonaventure University, Saint
Bonaventure 14778. *Study skin collection and noncredit course.*

Dr. Ronald Hartman, Director, Continuing Education, Saint Bonaventure
University, Saint Bonaventure 14778. *Noncredit course.*

Dr. W. W. Kanzler, Biology Department, Wagner College, Staten Island 10301.
Credit course.

Department of Ecology and Evolution, State University of New York at Stony
Brook, Stony Brook 11794. *Credit course.*

School of Continuing Education, State University of New York College of
Environmental Science and Forestry, Syracuse 13210. *Credit course.*

North Carolina

Hampton Mariner's Museum, 120 Turner Street, Beaufort 28516. *Noncredit course.*

Dr. J. F. Randall, Biology Department, Appalachian State University, Boone
28608. *Credit course.*

Dr. R. F. Soots, Biology Department, Campbell University, Buies Creek 27506.
Credit and noncredit courses.

Department of Zoology, Wilson Hall 046-A, University of North Carolina, Chapel Hill 27514. *Credit course.*

Dr. Richard Brown, Biology Department, University of North Carolina, Charlotte 28223. *Study skin collection and credit and noncredit courses.*

Central Piedmont Community College, P.O. Box 4009, Charlotte 28204. *Credit and noncredit courses.*

Dr. Jack H. Fehon, Department of Biology, Queens College, Charlotte 28274. *Credit course.*

R. C. Lindsay, Haywood Technical Institute, P.O. Box 457, Clyde 28721. *Credit course.*

Zoology Department, Duke University, Durham 27706. *Credit course.*

Agriculture Department, Wayne Community College, Caller Box 8002, Goldsboro 27530. *Credit course.*

Dr. L. Moseley, Biology Department, Guilford College, Greensboro 27410. *Credit course.*

Biology Department, East Carolina University, Greenville 27834. *Credit course.*

Chairman, Department of Biology, Pembroke State University, Pembroke 28372. *Credit course.*

Department of Zoology, North Carolina State University at Raleigh, Raleigh 27606. *Credit and noncredit courses.*

F. Burgin, Jr., Isothermal Community College, Box 804, Spindale 28960. *Credit course.*

Dr. J. Parnell, Biology Department, University of North Carolina at Wilmington, Wilmington 28403. *Credit course.*

North Dakota

Myron L. Freeman, Dickinson State College, Dickinson 58601. *Credit course.*

Dr. F. Cassel, Department of Biology, North Dakota State University, Fargo 58102. *Credit course.*

Dr. R. D. Crawford, Department of Biology, University of North Dakota, Grand Forks 58202. *Credit course.*

William L. Moore, Valley City State College, Valley City 58072. *Credit course.*

Ohio

Dr. F. S. Orcutt, Jr., Department of Biology, University of Akron, Akron 44325. *Credit course.*

C. N. Currier, Kent State University, Ashtabula Campus, 3325 W. 13th Street, Ashtabula 44004. *Noncredit course.*

Dr. E. W. Martin, Biological Sciences, 414 Life Science, Bowling Green State University, Bowling Green 43403. *Credit course.*

E. Beisbrock-Didham, Continuing Education, 238 Administration Building, Bowling Green State University, Bowling Green 43403. *Noncredit course.*

Biology Department, Malone College, 515 25th Street N.W., Canton 44709. *Credit course.*

Dr. J. L. Gottschang, Department of Biological Sciences, University of Cincinnati, Cincinnati 45221. *Credit course.*

Department of Zoology, Ohio State University, 1735 Neil Avenue, Columbus 43210. *Credit course.*

Division of Continuing Education, Ohio State University, 2400 Olentangy River Road, Columbus 43210. *Noncredit course.*

Dr. E. H. Burtt, Jr., Department of Zoology, Ohio Wesleyan University, Delaware 43015. *Credit course.*

Cincinnati Nature Center, 4949 Tealtown Road, Milford 45150. *Noncredit course.*

Mount Vernon Nazarene College, Martinsburg Road, Mount Vernon 43050. *Credit course.*

Hocking Technical College, Admissions Department, Nelsonville 45764. *Credit course.*

Dr. D. Osborne, Zoology Department, Miami University, Oxford 45056. *Credit course.*

Dr. Louis Laux, Jr., Wittenberg University, Springfield 45501. *Credit and noncredit courses.*

Biology Department, Heidelberg College, Tiffin 44883. *Credit course.*

Dr. E. B. McLean, Department of Biology, John Carroll University, University Heights 44118. *Credit course.*

Dr. J. Willis, Life Science Department, Otterbein College, Westerville 43081. *Credit and noncredit courses.*

SMORGASBIRD, Antioch Outdoor Education Center, 1075 Route 343, Yellow Springs 45387. *Noncredit course.*

Biology Department, Antioch College, Yellow Springs 45387. *Credit course.*

Oklahoma

Dr. W. A. Carter, Biology Department, East Central State College, Ada 74820. *Credit and noncredit courses.*

Dr. D. A. Shorter, Department of Biology, Northwestern State University, Alva 73717. *Credit course.*

Dr. Charles M. Mather, Department of Science and Math, University of Science and Arts of Oklahoma, Chickasha 73018. *Noncredit course.*

Dr. W. J. Radke, Biology Department, Central State University, Edmond 73034. *Study skin collection and credit and noncredit courses.*

A. Marguerite Baumgartner, Box 51A, Route 2, Jay 74346. *Noncredit course.*

Department of Zoology, University of Oklahoma, Norman 73019. *Credit course.*

Oklahoma Center for Continuing Education, University of Oklahoma, Norman 73018. *Noncredit course.*

Director, School of Biological Sciences, Oklahoma State University, Stillwater 74074. *Credit and noncredit courses.*

E. Grigsby, Division of Natural Science and Math, Northeastern Oklahoma State University, Tahlequah 74464. *Credit course.*

Oregon

Biology Department, Linn-Benton Community College, Albany 97321.
Credit and noncredit courses.

Department of Continuing Education, Central Oregon Community College,
Bend 97701. *Noncredit course.*

Science and Math Department, Central Oregon Community College, Bend
97701. *Credit and noncredit courses.*

Dr. D. Ferguson, Malheur Field Station, P.O. Box 989, Burns 97720.
Credit and noncredit courses.

Southwestern Oregon Community College, Coos Bay 97420. *Credit and
noncredit courses.*

Herb Wisner, Biology Department, University of Oregon, Eugene 97403.
Study skin collection and noncredit course.

L. W. Spring, Department of Natural Science/Mathematics, Oregon College of
Education, Monmouth 97361. *Credit course.*

Department of Biology, George Fox College, Newberg 97132. *Credit course.*

Biology Department, Clackamas Community College, 19600 S. Molalla Avenue,
Oregon City 97045. *Noncredit course.*

Pennsylvania

Biology Department, Muhlenberg College, Allentown 18104. *Credit course.*

R. G. Sagar, Department of Biological Sciences, Bloomsburg State College,
Bloomsburg 17815. *Credit and noncredit courses.*

J. E. Williams, Department of Biology, Clarion State College, Clarion 16214.
Credit course.

Dr. R. L. Rymon, Department of Biology, East Stroudsburg State College, East
Stroudsburg 18301. *Credit course.*

Dr. D. Snyder, Biology Department, Edinboro State College, Edinboro 16444.
Credit course.

Natural Sciences Department, Messiah College, Grantham 17027. *Credit and
noncredit courses.*

John J. Ford, Harrisburg Area Community College, 3300 Cameron Street,
Harrisburg 17810. *Noncredit course.*

Department of Biology, Indiana University of Pennsylvania, Indiana 15701.
Credit course.

Division of Natural Sciences, 116 E & S Building, University of Pittsburgh,
Johnstown 15904. *Credit and noncredit courses.*

School of Continuing Education, Kutztown State College, Kutztown 19530.
Credit course.

Dr. J. Harclerode, Department of Biology, Bucknell University, Lewisburg
17837. *Credit course.*

Dr. P. W. Schwalbe, Department of Biological Sciences, Lock Haven State
College, Lock Haven 17745. *Credit course.*

Prof. K. G. Miller, Department of Biology, Millersville State College, Millersville 17551. *Credit course.*

Eastern College, St. Davids 19087. *Credit course.*

Dr. M. Carey, Department of Biology, University of Scranton, Scranton 18510. *Credit course.*

Dr. G. Turdik, Department of Biology, Slippery Rock State College, Slippery Rock 16057. *Credit course.*

SCASD Continuing Education, 131 W. Nittany Avenue, State College 16801. *Noncredit course.*

Dr. T. Williams, Biology Department, Swarthmore College, Swarthmore 19081. *Credit course.*

Pennsylvania State University, University Park 16802. *Credit course.*

Mr. H. G. Jones, Department of Biology, West Chester State College, West Chester 19380. *Credit course.*

Biology Department, Lycoming College, Williamsport 17701. *Credit course.*

Ms. Toni Greggerson, Westmoreland County Community College, Youngwood 15697. *Noncredit course.*

Rhode Island

Department of Zoology, University of Rhode Island, Kingston 02881. *Credit course.*

South Carolina

Maj. D. M. Forsythe, Department of Biology, The Citadel, Charleston 29409. *Credit course.*

Biology Department, College of Charleston, Charleston 29401. *Credit course.*

Department of Zoology, Clemson University, Clemson 29631. *Credit course.*

Dr. W. P. Pielou, Biology Department, Furman University, Greenville 29613. *Credit course.*

Richard Fox, Biology Department, Lander College, Greenwood 29646. *Noncredit course.*

J. W. Glenn, Jr., Biology Department, University of South Carolina—Lancaster, Lancaster 29720. *Noncredit course.*

South Dakota

Dr. D. Tallman, Department of Natural Science, Northern State College, Aberdeen 57401. *Credit course.*

Chairman, Biology Department, Dakota State College, Madison 57042. *Credit course.*

Dr. James E. Martin, Black Hills Natural Science Field Station, South Dakota School of Mines and Technology, Rapid City 57701. *Credit course.*

Dr. G. W. Blankespoor, Department of Biology, Augustana College, Sioux Falls 57102. *Credit course.*

Tennessee

Department of Biology, University of Tennessee, Chattanooga 37377. *Credit course.*

Dr. J. R. Freeman, Biology Department, University of Tennessee, Chattanooga 37401. *Noncredit course.*

Department of Biology, Austin Peay State University, Clarksville 37040. *Credit course.*

Mr. E. O. Grundset, Biology Department, Southern Missionary College, Collegedale 37315. *Credit course.*

Chairman, Department of Biology, Box 5063, Tennessee Technological University, Cookeville 38501. *Credit course.*

Division of Extended Services, Tennessee Technological University, Cookeville 38501. *Noncredit course.*

Biology Department, Lincoln Memorial University, Harrogate 37752. *Credit course.*

Department of Biology, Lambuth College, Jackson 38301. *Credit course.*

Dr. F. Alsop, Department of Biological Sciences, Kingsport University Center, East Tennessee State University, University Boulevard, Kingsport 37660. *Credit course.*

Department of Biological Sciences, University of Tennessee, Martin 38238. *Credit course.*

R. E. Van Cleef, Memphis P. P. Museum, 3050 Central, Memphis 38111. *Noncredit course.*

Department of Biology, Memphis State University, Memphis 38152. *Credit course.*

Department of Biology, Middle Tennessee State University, Murfreesboro 37132. *Credit course.*

Texas

Richard O. Albert, 310 N. Wright Street, Alice 78332. *Noncredit course.*

Dr. A. M. Powell, Sul Ross State University, Box C-64, Alpine 79830. *Credit course.*

Dr. Robert Neill, Department of Biology, University of Texas at Arlington, Arlington 76019. *Credit course.*

Brother Daniel Lynch, CSC, St. Edwards University, Austin 78704. *Credit course.*

T. H. Hamilton or R. F. Martin, Department of Zoology, University of Texas at Austin, Austin 78712. *Credit course.*

Informal Class Program, Texas Union, University of Texas at Austin, Austin 78712. *Noncredit course.*

Dr. Jed J. Ramsey, Lamar University, Box 10037, Lamar Station, Beaumont 77710. *Credit course.*

S. D. Castro, Department of Biology, Mary Hardin-Baylor College, Belton 76513. *Credit course.*

A. M. Pullen, Department of Biology, East Texas State University, Commerce 75428. *Credit course.*

Dr. B. R. Chapman, Division of Biology, University of Corpus Christi, Corpus Christi 76412. *Credit course.*

Dr. V. Allison, Department of Biology, Southern Methodist University, Dallas 75275. *Credit course.*

Steve Runnels, Curator of Ornithology, Dallas Museum of Natural History, P.O. Box 26193, Fair Park Station, Dallas 75226. *Study skin collection and noncredit course.*

Dr. P. James, Department of Biology, Pan American University, Edinburg 78539. *Credit course.*

Bill Voss, Museum of Science and History, 1501 Montgomery, Fort Worth 76107. *Study skin collection and noncredit course.*

Dr. R. R. Moldenhauer, Department of Life Sciences, Sam Houston State University, Huntsville 77341. *Credit course.*

W. Pulich, Fort Worth Audubon Society, 2021 Rosebud, Irving 75060. *Noncredit course.*

Dr. M. K. Rylander, Department of Biological Sciences, Texas Tech University, Lubbock 79409. *Credit course.*

Department of Biology, SFA Box 3003, Stephen F. Austin State University, Nacogdoches 79562. *Credit course.*

Director, Llano Estacado Museum, Wayland College, Plainview 79072. *Noncredit course.*

Dr. J. Rutledge, Division of Life Sciences, University of Texas at San Antonio, San Antonio 78285. *Credit course.*

Dr. D. Huffman, Biology Department, Southwest Texas State University, San Marcos 78666. *Credit and noncredit courses.*

Department of Biology, University of Texas at Tyler, Tyler 75701. *Credit course.*

Utah

Department of Biology, UMC 53, Utah State University, Logan 84322. *Credit course.*

Dr. C. D. Marti, Department of Zoology, Weber State College, Ogden 84408. *Credit course.*

Vermont

Wildlife Biology Program, University of Vermont, Burlington 05405. *Credit course.*

Dr. D. H. Miller, Department of Science, Lyndon State College, Lyndonville 05851. *Credit course.*

Prof. William Barnard, Cabot Science, Norwich University, Northfield 05663. *Credit course.*

Dr. Bob Jervis, Goddard College, Plainfield 05667. *Credit course.*

Vermont Institute of Natural Science, Woodstock 05091. *Noncredit course.*

Virginia

Dr. C. S. Adkisson, Biology Department, Virginia Polytechnic Institute, Blacksburg 24061. *Credit course.*

Department of Biology, Virginia Intermont College, Bristol 24201. *Credit course.*

D. S. Lancaster Community College, Rt. 60 W., Clifton Forge 24422.
  Credit and noncredit courses.
Dr. C. H. Ernst, Department of Biology, George Mason University, Fairfax
  22030. *Credit course.*
Ronald Stephens, Box 36, Ferrum College, Ferrum 24088. *Credit course.*
Dr. J. E. Fisher, Department of Biology, James Madison University,
  Harrisonburg 22807. *Credit course.*
Dr. A. C. Mellinger, Department of Biology, Eastern Mennonite College,
  Harrisonburg 22801. *Credit course.*
Department of Biology, Randolph-Macon Woman's College, Lynchburg 24503.
  Credit course.
Prof. R. Simpson, Department of Biology, Lord Fairfax Community College,
  Middletown 22645. *Study skin collection and credit course.*
W. Berghaus, Office of Continuing Education, Lord Fairfax Community
  College, Middletown 22645. *Noncredit course.*
Dr. R. Rose, Department of Biological Sciences, Old Dominion University,
  Norfolk 23508. *Noncredit course.*
Cornelis Laban, Richard Bland College, Petersburg 23803. *Credit and
  noncredit courses.*
J. Sargent, Reynolds Community College, P.O. Box 12084, Richmond 23241.
  Credit course.
Admissions, Virginia Commonwealth University, Richmond 23284. *Credit course.*
Continuing Education, Virginia Commonwealth University, Richmond 23284.
  Noncredit course.
John F. Mehner, Ph.D., Mary Baldwin College, Staunton 24401. *Credit course.*
Department of Biology, Sweet Briar College, Sweet Briar 24595. *Credit course.*
Special Ecology Programs, Sweet Briar College, Sweet Briar 24595.
  Noncredit course.
Drs. M. Bird and R. Beck, Department of Biology, College of William and
  Mary, Williamsburg 23185. *Credit and noncredit courses.*

Washington
Office of Continuing Education, Grays Harbor College, Aberdeen 98520.
  Noncredit course.
Dennis White, Biology Department, Green River Community College, Auburn
  98002. *Noncredit course.*
Department of Biology, Western Washington University, Bellingham 98225.
  Credit course.
Dr. J. Johns, Department of Biology, Eastern Washington University, Cheney
  99004. *Credit and noncredit courses.*
S. van Niel, Biology Department, Everett Community College, Everett 98201.
  Noncredit course.
Dr. S. G. Herman, Evergreen State College, Olympia 98505. *Credit course.*

Department of Zoology, Washington State University, Pullman 99164. *Credit course.*

Dr. D. L. Hicks, Biology Department, Whitworth College, Spokane 99251. *Credit course.*

Chairman, Department of Biology, Pacific Lutheran University, Tacoma 98447. *Credit course.*

J. Shelley, Fort Steilacoom Community College, 9401 Farwest Drive, Tacoma 98498. *Credit and noncredit courses.*

West Virginia

Dr. L. E. Bayless, Department of Biology, Concord College, Athens 24712. *Credit course.*

Dr. A. R. Buckelew, Jr., Biology Department, Bethany College, Bethany 26032. *Credit course.*

A. F. Michna, Fairmont State College, Fairmont 26554. *Credit course.*

Department of Biological Sciences, Marshall University, Huntington 25701. *Credit course.*

Dr. Thomas K. Pauley, Department of Natural Science, Salem College, Salem 26426. *Credit course.*

Wisconsin

Dr. R. L. Verch, Biology Department, Northland College, Ashland 54806. *Credit course.*

Dr. J. Lutz, Department of Biology, Beloit College, Beloit 53511. *Credit course.*

Lakeshore Technical Institute, 1290 North Avenue, Cleveland 53015. *Noncredit course.*

Dr. O. Owen, Department of Biology, University of Wisconsin—Eau Claire 54701. *Credit and noncredit courses.*

Registrar, Marian College of Fond du Lac, 45 S. National Avenue, Fond du Lac 54935. *Credit and noncredit courses.*

Prof. R. Stiehl, ES 317, University of Wisconsin—Green Bay, Green Bay 54302. *Credit course.*

M. Engelman, Director of Outreach, LC 803, University of Wisconsin—Green Bay, Green Bay 54302. *Noncredit course.*

Chairman, Biology Department, Blackhawk Technical Institute, 2228 Center, Janesville 53545. *Noncredit course.*

Dr. K. Baugrud, University Extension, University of Wisconsin—Parkside, Kenosha 53141. *Noncredit course.*

Biology Department, University of Wisconsin—La Crosse, La Crosse 54601. *Credit course.*

Department of Biology, Mount Scenario College, Ladysmith 54848. *Credit course.*

Chairperson, Biology Department, Edgewood College, 855 Woodrow Street, Madison 53711. *Credit course.*

Continuing Education, Edgewood College, 855 Woodrow Street, Madison 53711. *Noncredit course.*

Sister Julia Van Denack, Silver Lake College of the Holy Family, Manitowoc 54220. *Credit and noncredit courses.*

Dr. R. H. Wilson, Biology Department, University of Wisconsin—Stout, Menomonie 54751. *Credit course.*

Zoology Department, University of Wisconsin—Milwaukee, Milwaukee 53201. *Credit course.*

Department of Biology, University of Wisconsin—Platteville, Platteville 53818. *Credit course.*

Dr. M. Gage, Nicolet College, Rhinelander 54501. *Noncredit course.*

W. Brooks, Chairman, Department of Biology, Ripon College, Ripon 54971. *Credit course.*

Director, Pigeon Lake Field Station, University of Wisconsin—River Falls, River Falls 54022. *Credit course.*

Biology Department, University of Wisconsin—Stevens Point, Stevens Point 54481. *Credit course.*

Chairman, Department of Biology, University of Wisconsin—Superior, Superior 54880. *Credit course.*

Dr. C. North, Biology Department, University of Wisconsin—Whitewater, Whitewater 53190. *Credit and noncredit courses.*

Wyoming
Department of Zoology, University of Wyoming, Laramie 82071. *Credit and noncredit courses.*

Puerto Rico
Department of Biology, Cayey University College, Cayey 00633. *Credit course.*
Department of Biology, University of Puerto Rico, Mayagüez 00708. *Credit course.*
Biology Department, Antillian College, Box 118, Mayagüez 00708. *Credit course.*

West Indies
Asa Wright Nature Center, P.O. Bag 10 P.O.S., Trinidad, *or* Wonder Bird Tours, 500 Fifth Avenue, New York, N.Y. 10036. *Noncredit course.*

CANADA
Alberta
Registrar, University of Calgary, Calgary T2N 1N4. *Credit course.*
The Head, Department of Zoology, University of Alberta, Edmonton T6G 2E9. *Credit course.*

British Columbia
Chairman, Department of Biological Sciences, Simon Fraser University, Burnaby V5A 1S6. *Credit course.*

Manitoba

Office of Continuing Education Summer Session, Brandon University, Brandon R7A 6A9. *Noncredit course.*

Dr. S. G. Seally, Department of Zoology, University of Manitoba, Winnipeg R3T 2N2. *Credit course.*

New Brunswick

Dr. Dan Keppie, Department of Biology, University of New Brunswick, Fredericton E3B 5A3. *Credit course.*

Department of Biology, Mount Allison University, Sackville E0A 3C0. *Credit course.*

Newfoundland

Department of Biology, Memorial University of Newfoundland, St. John's A1B 3X9. *Credit course.*

Nova Scotia

Dr. P. C. Smith, Biology Department, Acadia University, Wolfville B0P 1X0. *Credit course.*

Ontario

Outdoor Art and Science School, P.O. Box 112, Guelph N1H 6J6. *Noncredit course.*

Department of Zoology, University of Guelph, Guelph N1G 2W1. *Credit course.*

Department of Zoology, University of Western Ontario, London N6A 5B7. *Credit course.*

Department of Biology, University of Ottawa, Ottawa K1N 6N5. *Credit course.*

Department of Biology, Trent University, Peterborough K9J 7B8. *Credit course.*

J. P. Ryder, Biology Department, Lakehead UNW, Thunder Bay P7B 5E1. *Credit course.*

Department of Zoology, University of Toronto, Toronto M5S 1A1. *Credit course.*

Prince Edward Island

Department of Biology, University of Prince Edward Island, Charlottetown C1A 4P3. *Credit course.*

Quebec

Les Jeunes Explos, CP 10, St.-Fulgence, Chicoutimi G0V 150. *Noncredit course.*

Centre Ecologique de Port ou Saumon, St.-Fidele, Conti Charlevaux G0T 1T0. *Noncredit course.*

Conseil de la Jeunesse Scientifique, 1415 est Jarry, Montreal H2E 2Z7. *Noncredit course.*

Centre Ecologique de Port ou Saumon, College Bourget, CP 1000, Rigaud G0P 1P0. *Noncredit course.*

Dr. Andre Cyr, Biology Department, Faculty of Science, Sherbrook University, Sherbrook J1K ZR1. *Credit course.*

Saskatchewan

Dr. D. M. Secoy, Department of Biology, University of Regina, Regina S4S 0A2.
  Credit course.
Department of Biology, University of Saskatchewan, Saskatoon S7N 0W0.
  Credit course.

INTERNATIONAL BIRD TOURS

Organized tours can take adventurous bird enthusiasts to nearly every habitat on earth, providing excellent opportunities to compare North American birds and habitats with those of other countries. Under the supervision of experienced leaders, participants may have intimate contact with a remarkable variety of birds. Even such unlikely named birds as glittering-bellied emeralds and paltry tyrannulets may become familiar in the distant lands visited by bird tour groups. Antarctic ice floes, volcano interiors, and remote jungle rivers are just a few of the exotic habitats within reach of international bird tour participants.

In addition to providing memorable bird-watching experiences, such tours also serve the vital function of directing attention and finances toward important wildlife habitats. Wildlife-oriented tourism is an increasingly important industry for many developing nations because the tours are demonstrating that wildlife and wild lands can generate more dollars from tourism than from lumber, fur, and feathers.

The fifty-two sponsors that are given in the following lists offer organized bird-watching tours to one or more of ninety-six countries and regions. The tours are at least one week long and depart throughout the year, depending on the season and climate of the country to be visited. Locales appearing in parentheses are either popular bird-watching areas within the region or former names for countries, e.g., Sri Lanka (Ceylon). The key to bird tour sponsors and their addresses follows the maps. Contact the sponsors for current trip costs and detailed itineraries.

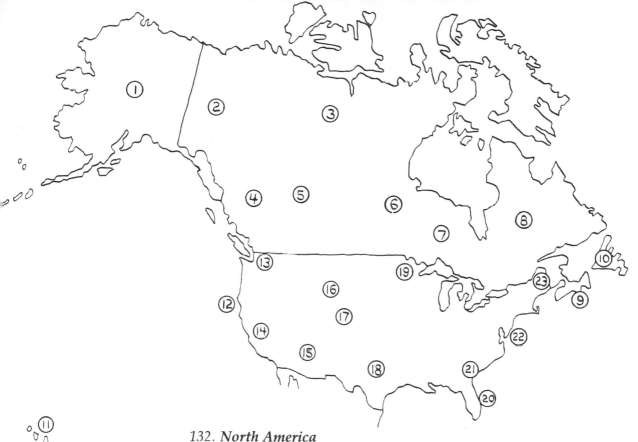

132. North America

1. Alaska (Pribilof Islands): AT, AW, BB, CN, DT, MA, MT, NA, NE, NY, PG, QU, VE, WI
2. Yukon Territory: QU
3. Northwest Territories: CA, MA, NA, NT, QU
4. British Columbia: NT, OR, QU, ST
5. Alberta: AU
6. Manitoba (Churchill): MA, NT, PG, QU, VE, WI
7. Ontario (Point Pelee): ME, NT, PG
8. Quebec (Bonaventure Island): MB, WI
9. Nova Scotia: MA, ME, RM, VE, WI
10. Newfoundland: MA, NT, RM, VE, WI
11. Hawaiian Islands: AW, NA, NT, NY, OS, QU, RM, WI
12. Pacific Coast: HM, ME, OS
13. Washington: MA, ST, WI
14. California: AW, BB, HM, MA, MT, OR, PR, RM, WI
15. Arizona: AU, AW, BB, MA, ME, NT, OR, PG, PR, RM, VE, WI
16. Wyoming: CN
17. Colorado: MA, VE, WI
18. Texas: BB, CD, MA, ME, NA, NT, PG, VE, WI, WN
19. Minnesota: BB
20. Florida (Everglades): BB, LF, MA, NA, NT, OR, PE, QU, VE, WI, WS
21. Georgia (Okefenokee): CN, WS
22. Atlantic Coast: OS
23. Maine: JV, MA, MB, ME, WI

133. Central America and the West Indies

1. Baja California: HM, MA, NE, NT, NY, QU, WI
2. Mexico: AU, AW, BB, CD, GS, IT, MA, ME, NE, PG, PR, RM, QU, VE, WI, WN
3. Yucatan: HT, IT, ME, NE, NT, QU, VE, WI
4. Belize: QU
5. Guatemela: IT, MA, ME, NA, PR, QU, RM, UN, VE, WS
6. El Salvador: RM
7. Honduras: QU
8. Costa Rica: BB, HT, MA, MT, QU, PR, RM, SA, VE, WI
9. Panama: BB, HT, MA, ME, PG, QU, VE
10. West Indies: NT, RM, WL
11. Trinidad and Tobago: MA, NA, NT, NY, QU, RM, WB, WN
12. Jamaica: BB, VE
13. Puerto Rico: BB
14. Cayman Islands: RM
15. Cuba: BB, CT, MA, MT
16. Bahamas: RM

134. *South America and Antarctica*

1. Colombia: BB, HT, RM, VE, WL
2. Ecuador: MT, NY, QU, PR, SA, VE, WI, WN
3. Galapagos Islands: HE, LT, MT, NA, NE, NT, NY, OS, QU, SA, SE, VE, WI
4. Peru: AN, BB, CN, HE, MA, MT, PR, QU, SA, VE
5. Bolivia: VE
6. Chile: MT, WL
7. Antarctica: LT, NY, SE
8. South Georgia: LT
9. Argentina (Patagonia and Tierra del Fuego): BB, HT, MT, NY, SA, SE, WL
10. Falkland Islands: LT, MT, NY, QU, SA
11. Brazil (Amazon): BB, HT, NY, QU, SA, SE, VE, WL
12. Surinam: BB, VE, WB, WI
13. Venezuela: BB, MA, VE

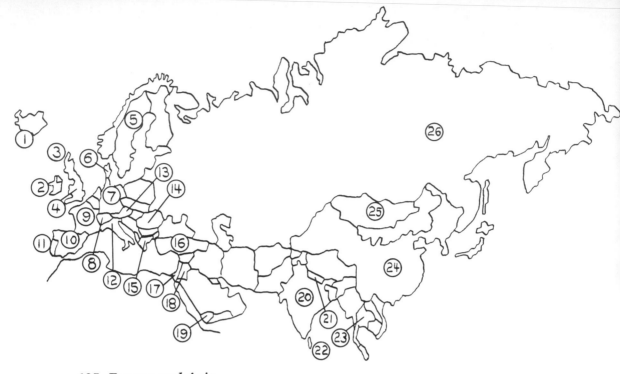

135. *Europe and Asia*

1. Iceland: NY, SE, QU
2. Ireland: NA, OS
3. Scotland (Hebrides): BB, HE, NA, NE, NT, NY, OS, QU, WI
4. England: BB, BG, MT, NA, NT, OS, SH, WI
5. Sweden: MA, NA, SH
6. Denmark: MA, NA, SH
7. Germany: NA
8. Switzerland: MA, NA, QU, RM
9. France: NA, SH, WI
10. Spain: MA, MT, NA, NJ, QU, SH
11. Portugal: MA
12. Italy: NJ
13. Austria: NJ
14. Romania: SH
15. Greece: NA, NJ, SH, QU
16. Turkey: MA, SH, WI
17. Israel: BB, MA, NA, SH
18. Jordan: SH
19. Yemen: SH
20. India (Kashmir, Ladakh): BB, BK, ET, HE, MA, MT, NA, NH, NY, QU, RM, SH, WN
21. Nepal (Himalayas, Bhutan): BB, BK, HE, LT, MA, MT, NA, NH, NY, QU, SH, WI, WN
22. Sri Lanka (Ceylon): BB, BK, HT, NY, QU, SH
23. Thailand: BB, WI
24. China: AW, BB, MA, NY, WL
25. Mongolia: WL
26. Siberia: SH, WI

136. *Africa*

1. Morocco: SH, WI
2. Senegal: WL
3. Cameroons: NY, VE
4. Botswana: NY, VW, WL
5. Zambia: ET
6. Malawi: NY
7. Zaire (Belgian Congo): QU, WL
8. Rwanda: QU
9. Tanzania: BB, ES, HT, MT, NT, NE, NY, QU
10. Kenya: BB, ET, HT, MA, MT, NA, NE, NT, NY, PG, QU, SH, VE, WI, WN
11. Seychelles: ET, NE, NT, NY, SE, SH, WA
12. Madagascar: NY, WI

137. *Australasia*

1. Philippines: WL
2. Indonesia (Singapore, Malaysia, Borneo, Sumatra): BB, HE, QU, WL
3. New Guinea: DT, ES, NE, WI, WL
4. Australia: BB, DT, ES, FL, MA, NA, NE, NT, NY, PE, QU, WL
5. New Zealand: FL, MA, NA, NE, NY, PE, QU, WL
6. Marquesas Islands (Polynesia): HE, NA, OS
7. Mariana Islands, Guam, Caroline Islands, and Marshall Islands: WN

Key to Bird Tour Sponsors

AN Andean Odyssey Tours, P.O. Box 5540, Berkeley, Calif. 94705

AT Attour, 4333 N. Kedvale Avenue, Chicago, Ill. 60641

AU Audubon Ecology Field Seminars, National Audubon Society, P.O. Box 3557, Boulder, Colo. 80307

AW Audubon West, National Audubon Society, Western Education Center, 376 Greenwood Beach Road, Tiburon, Calif. 94920

BB Bird Bonanzas, Inc., 12550 Biscayne Boulevard, Suite 501, North Miami, Fla. 33181

BG Birdguide, Ashville, Rose Bank, Burley-in-Wharfdale, Ilkley, West Yorkshire LS29 7PG, England

CA Canoe Arctic, Inc., 9 John Beck Crescent, Brampton, Ontario LGW 2T2, Canada

CD Chihuahuan Desert Research Institute, Director of Education, P.O. Box 1334, Alpine, Tex. 79830

CN Cincinnati Nature Center, 4949 Tealtown Road, Milford, Ohio 45150

CT Cuba Travel, USA. P.O. Box 60394, Houston, Tex. 77205

DT Discovery Tours, American Museum of Natural History, Central Park West at 79th Street, New York, N.Y. 10024

ES Ecological Study Groups Birder's Safaris-African Wildlife, Dr. Jim C. Hitchcock, P.O. Box 1288, Nevada City, Calif. 95959

ET Esplanade Tours/Swans, 38 Newbury Street, Boston, Mass. 02116

GS Gorgas Science Society, c/o Fred Webster, 4926 Strass Drive, Austin, Tex. 78731

HE Hanns Ebensten Travel, Inc., 55 W. 42nd Street, New York, N.Y. 10036

HM H&M Landing, 2803 Emerson Street, San Diego, Calif. 92106

HT Holbrook Travel, Inc., 3520 N.W. 13th Street, Gainesville, Fla. 32601

IH Island Holidays, 214 Grant Avenue, San Francisco, Calif. 94108

IT International Training Programs, University of Oklahoma, 1700 Asp, Norman, Okla. 73037

JV Joseph Van Os Nature Tours, Box 655, Vashon, Wash. 98070

KB King Bird Tours, P.O. Box 196 Planetarium Station, New York, N.Y. 10024

LF Lykes Florida Wilderness Tours, Lykes Brothers, Inc., Division of Forestry and Recreation, P.O. Box 66, Palmdale, Fla. 33944

LT Lindblad Travel, Inc., 133 E. 55th Street, New York, N.Y. 10022

MA Massachusetts Audubon Society, Natural History Tours, Lincoln, Mass. 01773

ME Merlin Birding Tours, 1736 Albans, Houston, Tex. 77005

MP Monterey Bay Pelagic Excursions, c/o Debi Love Millichap, P.O. Box 842, Pacific Grove, Calif. 93950

MT Mountain Travel, 1398 Solano Avenue, Albany, Calif. 94706

NA National Audubon Society Tours™, 40 Hungerford Street, Hartford, Conn. 06106. Tours operated by Jornee Unlimited, Inc., under license by the National Audubon Society

NE Nature Expeditions International, 599 College Avenue, Palo Alto, Calif. 94306

NH Nature Himalayas, 2726 Mockingbird, Kalamazoo, Mich. 49008

NJ New Jersey Audubon Society. Tours operated by Mohawk Travel, 60 E. Mt. Pleasant Avenue, Livingston, N.J. 07039

NT Nature Travel Service (formerly Ontario Nature Tours), 6372 Montrose Road, Niagara Falls, Ontario L2H 1L6, Canada

NY New York Zoological Society, P.O. Box 108, Bronx, N.Y. 10460, Attention Iris Freed

OR Ornitholidays (Regd), 1/3 Victoria Drive, Bognor Regis, West Sussex P021 2PW, England

OS Oceanic Society Expeditions, Fort Mason, Building 315, San Francisco, Calif. 94123

PE Pacific Exploration Company, Box 3042, Santa Barbara, Calif. 93105

PG Peregrine, inc., P.O. Box 3062, Houston, Tex. 77001

PR Point Reyes Bird Observatory, 4990 Shoreline Highway, Stinson Beach, Calif. 94970

QU Questers Tours and Travel, Inc., 257 Park Avenue S., New York, N.Y. 10010

RM Russ Mason's Natural History Tours (formerly Flying Carpet Tours), P.O. Box Q, Kissimmee, Fla. 32741

SA South American Wilderness Adventures, Inc., 1760 Solano Avenue, Berkeley, Calif. 94707

SE Society Expeditions, 723 Broadway E., Seattle, Wash. 98102

SH Sunbird Holidays, Executive Travel (Holidays), Ltd., 141 Sloane Street, London SW1X 9BJ, England

ST Swiftsure Tours, Ltd., Michael G. Shepard, Tour Director, 119–645 Fort Street, Victory, British Columbia V8W 1G2, Canada

UN Unitrex, 1043 E. Green Street, Pasadena, Calif. 91106

VE Victor Emanuel Nature Tours, Inc., P.O. Box 4429, Austin, Tex. 78765

WB Wonder Bird Tours, 500 Fifth Ave., New York, N.Y. 10036

WI Wings Inc. (formerly Northeast Birding and Will Russell Weekends), Box 287, Seal Harbor, Maine 04675

WL Wanderlust, c/o Peter Alden, Massachusetts Audubon Society, Lincoln, Mass. 01773

WN World Nature Tours, Inc., P.O. Box 693, Silver Springs, Md. 20901

WS Wilderness Southeast, Rt. 3, Box 619, Savannah, Ga. 31406

WW Wilderness World, 1342 Jewell Avenue, Pacific Grove, Calif. 93950

6

RESEARCH PROGRAMS WELCOMING AMATEURS

Every Christmas season the National Audubon Society organizes more than thirty thousand bird-watchers to survey the North American winter bird population. More than three thousand participants assist the United States and Canadian wildlife services with the annual Breeding Bird Survey, and hundreds of observers participate in a count of bald eagles wintering on the Mississippi. There are also organized counts of backyard feeder birds, hawks, loons, woodcocks, nighthawks, cranes, waterfowl, and birds that frequent garbage dumps. Ambitious programs exist to locate range extensions for the endangered Kirtland's warbler and to assemble bluebird nest box "trails" to help reverse a serious population decline. Rock climbers carve nesting ledges into Alberta rock faces for cliff-nesting Canada geese, and volunteers wade through Idaho marshes to capture young herons for a banding study. These are only a few of the ways that amateur bird enthusiasts assist North American bird research programs.

The science of ornithology is fortunate to have a high proportion of amateur participants. Organizers of an increasing number of bird research programs recognize the growing number of willing amateur participants and welcome their assistance. Without volunteers, many of North America's most ambitious bird

138. The Breeding Bird Survey attempts to detect changes in North American land bird populations in order to identify population declines and to employ management techniques before a species becomes threatened or endangered.

studies would be inoperative. Benefits are derived from both directions. Professional ornithologists receive enthusiastic and often skilled assistance, and amateur bird-watchers learn more about bird biology and find satisfaction from assisting researchers.

The ranks of amateur ornithologists include a high proportion of professionals who offer a wide assortment of valuable experience and skills. Organizers of research programs should recognize that most amateurs watch birds during their leisure time for relaxation and recreation; therefore, the most successful cooperative programs are those that retain the "fun" element of bird-watching.

Bird counts, censuses, and surveys are the most frequent "employers" of amateur participants. Without such input, most large-scale population estimates and distribution studies would not generate enough field data to reveal continent-wide or local trends. Likewise, programs such as nest record card schemes and colonial bird inventories would be ineffective without the assistance of amateurs who collectively generate massive amounts of data. Although competent identification skills are a must for most of these programs, there is almost always room for beginners to help with record keeping and bird spotting.

A growing number of states and provinces are launching nongame research programs to benefit songbirds, raptors (hawks, owls, and vultures), and other long-neglected species. Such programs rely largely on amateurs to identify critical habitats and record wildlife distributions. Such input permits professionals to map the range of uncommon species and to design management plans to protect threatened habitats.

This chapter is a directory of bird research programs and was compiled from information submitted by project organizers throughout North America. Enthusiastic responses to a request for information were received from many North American bird clubs and wildlife agencies. As a result, the directory includes descriptions of most continent-wide and regional research programs that rely on assistance from amateur bird-watchers. The remarkable variety of programs available at the local level is listed under state and provincial headings. It is hoped that this survey of bird-watching activities will encourage even more amateur participation in ornithological research and will inspire new projects and further cooperation between amateurs and professionals.

BREEDING SEASON PROGRAMS

BREEDING BIRD SURVEY
Nongame Section, Migratory Bird and Habitat Research Laboratory, Laurel, Md. 20811, *or* Migratory Birds, Box 1590, Sackville, New Brunswick E0A 3C0, Canada

Sponsored by the U.S. Fish and Wildlife Service and the Canadian Wildlife Service, the Breeding Bird Survey attempts to detect changes in North American land bird populations in order to identify population declines and to employ management techniques before a species becomes threatened or endangered. Observers are requested to pick a day in June with good weather conditions on a date as close as possible to previous "runs." Each observer is assigned a randomly selected 25-mile-long route and begins exactly one-half hour before local sunrise, counting and recording all birds detected during three minutes at the starting point. The counting is repeated at forty-nine additional stops, each $\frac{1}{2}$ mile apart. Only birds counted during the fifty three-minute stops are included in the totals. A route should take from four to four-and-a-half hours to complete. Birds should not be coaxed with recorded sounds. It is important to finish in the time frame because on most mornings bird song decreases rapidly after the first four hours. Since these results are frequently used in policymaking and environmental impact assessments, it is extremely important that procedures be followed closely.

There are usually two people in each party—an observer and a record keeper. The routes are on secondary roads in order to minimize interference from traffic noise and danger to observers. Approximately eighteen hundred routes are surveyed every year. When the routes are completed, all field data are sent to the Nongame Section of the Migratory Bird and Habitat Research Laboratory in Maryland. After a thorough review by staff biologists, all data are transferred to magnetic tape and subjected to computer edit checks. Three listings are produced: one sorted by individual route; one by state and province; and one by species. These listings are available to the public, and the data are available for appropriate research. An annual newsletter is sent to all participants.

Participants should be able to identify birds by sight and sound in their area. Knowledge of bird songs and calls is the most crucial factor, since the short time spent at each stop means that most birds recorded are heard and not seen. Severe hearing deficiencies can render the results of a route unusable. Beginners may want to accompany a qualified observer to assist with note taking or driving. This is an excellent opportunity to learn bird songs from an experienced observer.

BREEDING BIRD CENSUS
The Editor, *American Birds,* National Audubon Society, 950 Third Avenue, New York, N.Y. 10022

Sponsored by the National Audubon Society, the census determines breeding bird composition and density in various types of habitats across North America. The censuses are conducted in relatively small (up to 200 acres) tracts of intensively surveyed and studied land through repeated visits during the breeding season. Volunteers can originate new studies and participate in ongoing studies. Activities include accurate mapping of the tract, tract descriptions, botanical surveys, and location and description of breeding bird populations. Observers write up their observations following a standardized format, and photographic talent is useful. The leader of each survey should be a trained biologist, and amateurs and students can help with plotting, botany, and bird location. The help of trained botanists or foresters is most useful. Instruction reprints are available, and the results of the census are published in the January issue of *American Birds.*

NORTH AMERICAN NEST RECORD CARD PROGRAM (NRCP)
Laboratory of Ornithology, Cornell University, 159 Sapsucker Woods Road, Ithaca, N.Y. 14850

The North American Nest Record Card Program, conducted under the auspices of the Laboratory of Ornithology and the National Audubon Society, collects, processes, and stores information on the nesting biology of birds. Hundreds of amateur and professional field ornithologists annually contribute thousands of

139. The Breeding Bird Census has contributed much information about expanding bird populations such as those of the red-winged blackbird. Photo by Leslie McKim

cards to the program's headquarters at the Laboratory of Ornithology. When a nest is found, an observer records such information as the name of the species, location, habitat, and reproductive history on special nest record cards, which are provided free to participants. The NRCP seeks both historical and current information on nests. Often information contained in field notes of past years is easily transferred to nest cards.

The NRCP has collected data since 1965. There are more than two hundred and fifty thousand nest records on file. When records are sent to the laboratory, they are edited and verified for accuracy and usability. For species with more than a thousand records on file, data are stored in a computerized memory for efficient retrieval and use. Data contained in the NRCP files are available to any qualified researcher, professional or amateur, interested in areas of avian reproductive biology, such as variation in breeding season, clutch size, and nesting success. Per-

140. *Special nest record cards for the North American Nest Record Card Program are provided free to participants.*

sons wishing to use data from the program must submit a brief outline of the proposed use of the material. Data are provided in the form of a computer print-out, computer tape, punched IBM card decks, or photocopies of the original data cards. Fees are based on the cost of producing the data.

Participants visit nests, fill out special cards, and return completed cards to the NRCP. The cards are designed for easy use—the front of the card provides space for general information on the nest and its location, the back has space for actual observations. All information contributed to the program is of value, but nests on which information is given for more than one visit during the breeding season are ultimately of the greatest use. Careful bird students, either amateur or professional, are welcome to participate. Cards are then returned either directly to Cornell or to one of more than 150 regional centers located throughout the country. All contributors receive a semiannual newsletter.

CANADIAN NEST RECORDS SCHEME

Newfoundland Nest Records Scheme, c/o J. Maunder, Newfoundland Museum, Natural History Section, St. John's, Newfoundland A1C 1G9

Maritimes Nest Records Scheme, c/o Canadian Wildlife Service, P.O. Box 1590, Sackville, New Brunswick, E0A 3C0

Quebec Nest Record Card Program, c/o Dr. H. Ouellet, National Museum of Natural Sciences, National Museums of Canada, Ottawa, Ontario K1A 0M8

Ontario Nest Records Scheme, c/o Dr. G. Peck, Department of Ornithology, Royal Ontario Museum, 100 Queen's Park, Toronto, Ontario M5S 2C6

141. The nesting activity of the dark-eyed junco is well-documented due to the Canadian Nest Records Schemes.
Photo by Leslie McKim

Prairie Nest Records Scheme (includes District of Mackenzie, Northwest Territories), c/o H. Copland, Manitoba Museum of Man and Nature, 190 Rupert Avenue, Winnipeg, Manitoba R3B DN2

British Columbia Nest Records Scheme (includes Yukon Territory), c/o W. Campbell, British Columbia Provincial Museum, Victoria, British Columbia V8W 1A1

The Canadian Nest Records Schemes assemble data for future reference on the nesting biology of Canadian birds. Participants enter data on standard nest record cards for all nests of which the contents can be determined and the species identified. At the end of the field season, records are submitted to the central file in the appropriate region. Data forms are provided free on request to interested participants, who should have the ability to identify most breeding birds and to accurately observe and record bird behavior and other ecological data, as outlined on the cards.

COLONIAL BIRD REGISTER
Laboratory of Ornithology, Cornell University, 159 Sapsucker Woods Road,
Ithaca, N.Y. 14850

Sponsored by the National Audubon Society and the Laboratory of Ornithology, the register follows and predicts population changes among North American colonial birds. It identifies important colonies within specific locales and establishes baseline data on breeding success and reproductive biology. It also identifies colonies that might be lost as a result of development, such as human construction and drainage programs, and aids in the preparation of environmental impact statements.

Participants in the register visit colonies and detail their locations, sizes, species composition, habitats, disturbance factors, and other information. In instances where colonies are of particular interest or concern, detailed information on the reproductive status, and hence the productivity of the colony, may be requested. The survey forms are designed to require a minimum amount of time for completion by the investigator, and are supplied free of charge. Participants are careful bird students, either amateur or professional. Check with the Colonial Bird Register and your local bird club to see if other investigators have already censused the colony you intend to investigate *before* you visit the colony. Also inquire about what precautions you should take so that human disturbance is not a detrimental factor to the colony.

KIRTLAND'S WARBLER CENSUS AND HABITAT INVENTORY
Dr. Lawrence A. Ryel, Chief, Office of Surveys and Statistical Services,
Michigan Department of Natural Resources, Box 30028, Lansing, Mich. 48909

Amateur ornithologists can assist the management program for the endangered Kirtland's warbler by inventorying suitable pine habitats outside of the currently occupied areas and in spot-checking suitable stands for the presence of warblers. To date, Kirtland's warblers have not been found nesting anywhere except in the jack-pine plains of northern Lower Michigan. Unmated singing males, however, were located in Ontario in 1977 and 1978, in Quebec in 1978, and in Wisconsin in 1978 and 1979. The possibility exists, therefore, that breeding colonies may occur elsewhere in the Great Lakes region.

Observers need to be familiar with various eastern bird songs, particularly the warblers. Their hearing should not be impaired since, at best, Kirtland's warblers can be heard only up to $\frac{1}{4}$ mile away. The ability to identify trees and shrubs is a must. Workers must be able to read maps and know enough about land surveys to be able to pinpoint the location of warblers and habitat when found. No reimbursement for time or travel expenses is possible, and all help is strictly voluntary.

NORTH AMERICAN WOODCOCK SURVEY
Dr. W. R. Whitman, Canadian Wildlife Service, Atlantic Region, Department of Environment, P.O. Box 1590, Sackville, New Brunswick E0A 3C0

142. *Kirtland's warblers have not been found nesting anywhere except in the jack-pine plains of northern Lower Michigan, although breeding colonies may exist elsewhere in the Great Lakes region.* Photo courtesy of the Michigan Department of Natural Resources

Conducted in the eastern United States and Canada, this survey establishes an index of the breeding woodcock population in North America by surveying on a random basis all habitat types so that the results reflect the status of the overall woodcock population. Once a year, during the spring, a survey is conducted at dusk by counting the number of different birds heard "peenting" during a two-minute period at each of the ten stops located 0.4 miles apart. Interested naturalists and hunters may assist in Canada; in the United States, the program is conducted by U.S. Fish and Wildlife Service personnel.

143. The program Monitoring the Status of Colonially Nesting Fish-Eating Birds of the Great Lakes investigates gulls, terns, and herons, the latter shown here.

MONITORING THE STATUS OF COLONIALLY NESTING FISH-EATING BIRDS OF THE GREAT LAKES

Hans Blokpoel, Canadian Wildlife Service, Ontario Region, 2721 Highway 31, Ottawa, Ontario K1G 3Z7

This program attempts to obtain a documented overview of the distribution and numbers of colonially nesting, fish-eating birds on the Canadian Great Lakes system, including the St. Marys, Niagara, and upper Saint Lawrence rivers. Participants provide the organizer with information on former and existing colonies of gulls, terns, and herons. Information should include location of the colony, date and time of visit, numbers (counted or estimated) or nests by species, and, where possible, further information on the colony (substrata, vegetation, predation, human disturbance, etc.). Participants should have the ability to identify gulls, terns, and herons. Contact the organizer before visiting these colonies to avoid unnecessary disturbances. Heron records should be sent to the Ontario Heronry Inventory (see listing on page 220), and they will be forwarded later to this organizer.

CANADIAN HERONRIES TOUR OF INSPECTION

Dr. Jean-Luc DesGranges, Canadian Wildlife Service, Department of the Environment, 2700 Laurier Boulevard, 4th floor, P.O. Box 10, 100 Ste.-Foy, Quebec G1V 4H5

144. *Selected heron nests are visited by volunteers from the Canadian Heronries Tour of Inspection.*

This tour determines the status and trends of great blue heron populations throughout Canada. Selected heronries are visited by volunteer observers about two weeks prior to fledging. The number of occupied and vacant nests, as well as the number of young herons in the nests, are recorded. Eggshell fragments found on the ground are also gathered. Serious naturalists and conservationists are welcome to participate.

BLUEBIRD CONSERVATION

North American Bluebird Society, Box 6295, Silver Spring, Md. 20906

Bluebirds are in jeopardy primarily because they cannot find enough nesting places. Dead trees and wooden fence posts that provided nesting cavities in earlier times are disappearing along with the bluebird's habitat of open field and orchards. Bluebirds must also compete for nesting sites with introduced house sparrows and starlings. These troubles, combined with pesticides, harsh winters, raccoon and snake predation, and competition from other native species, are believed to have reduced the eastern bluebird population by as much as 90 percent.

The North American Bluebird Society is dedicated to promoting interest and participation in a continent-wide effort to halt and, hopefully, reverse the long downward trend in the population of all three species of bluebird; to conducting research on the causes of the bluebird population decline, on the improved designs for bluebird nest boxes, and on methods for protecting nesting bluebirds from predators and competitors; and to disseminating research and other information concerning bluebird conservation through the society's quarterly periodical, *Sialia.*

Participants may construct, erect, and monitor bluebird nesting boxes in a suitable habitat; establish and maintain bluebird "trails" through individual or group efforts; maintain records of bluebird nesting activities; and experiment with new nest box designs and methods for controlling predators and competitors.

WINTER AND MIGRATION PROGRAMS

CHRISTMAS BIRD COUNT

The Editor, *American Birds,* National Audubon Society, 950 Third Avenue, New York, N.Y. 10022

Since its beginning in 1900, when the birds in twenty-five small areas were tallied, the Christmas Bird Count, sponsored by the National Audubon Society, has expanded to more than 1,275 counts throughout North America. The first count in 1900 attracted only twenty-seven bird-watchers, but the ranks of the count have swollen to more than thirty thousand participants. Such massive participation has resulted in tallies of more than 120 million birds in a single counting season. Data from the Christmas Count is stored in computers and is available to professionals and serious amateurs for bird distribution and population studies (see "Early Winter Range Mapping" and "Analytical and/or Derivative Studies," pages 176 and 183).

Christmas Bird Counts are conducted in all North American provinces and states, as well as all Central American countries, all West Indies countries, and the northern rim (Caribbean side) of South America. During one twenty-four-hour day, participants count as completely as possible all the birds within a 15-mile-diameter circle. There is no limit to the number of participants on a count. New counts are welcome, but they should be located at least 100 miles from the nearest other counts and have at least ten participants. Overlapping of count areas is not permitted. Participants should be reasonably proficient in bird identification or accompany someone who is. Practice in estimating numbers is desirable. Some stamina is required for certain field work. Bird-feeder watchers are welcome to participate. Accepted counts are published in the July issue of *American Birds.* Contact your local bird club or *American Birds* at the address above for detailed rules and the location of nearby counts.

THANKSGIVING BIRD COUNT
Dr. Ernest P. Edwards, Sweet Briar College, Sweet Briar, Va. 24595

Conducted by Sweet Briar College and the Lynchburg Bird Club, the Thanksgiving Bird Count determines the occurrence and relative abundance of birds around bird feeders throughout the United States in late autumn to make comparisons between years and areas. Participants watch feeders, birdbaths, and grain on the ground within a predetermined circle 15 feet in diameter and count the birds that enter the circle within any one-hour period on Thanksgiving Day. Participants should be able to identify any bird that comes to the count circle or be able to identify the common species and report others with a question mark. Contact the count for further information and for copies of the special reporting form.

ANNUAL BIRD FEEDER CENSUS
Bird Feeders Society, P.O. Box 243, Mystic, Conn. 06355

This census surveys the status of feeder-bird populations throughout the United States once each year in February. Participants identify and count birds in the immediate vicinity of their backyard feeders and report the results. Findings are reported in the society's quarterly publication, *Around the Bird Feeder.* Participants must be able to make accurate bird identifications, and membership in the society is encouraged.

MIDWINTER ONE-DAY BALD EAGLE COUNT
Elton Fawks, 510 Island Avenue, East Moline, Ill. 62144

Sponsored by the National Wildlife Federation, this program counts the number of bald eagles, distinguishing adults from immature eagles, along the Mississippi River, from its source through Minnesota and Wisconsin. Eagles are also counted along the Mississippi River borders of Iowa, Illinois, and Missouri, and statewide counts are conducted in Kentucky and Tennessee. The objective is to survey the Mississippi River drainage to estimate the wintering bald eagle population. Participants with the ability to distinguish bald eagles from other large birds are welcome to participate.

WINTER BIRD POPULATION STUDY

The Editor, *American Birds,* National Audubon Society, 950 Third Avenue, New York, N.Y. 10022

The Winter Bird Population Study intensively investigates winter bird populations of small tracts of land or tracts of land and water to determine species composition and density. These studies entail detailed plot measurements, floristic composition, and repeated visits during "winter" months in North America. Often the same tracts studied in breeding seasons are investigated. Volunteers originate new studies or participate in ongoing research. Activities include plot charting, floristic analysis and plant identification, bird censusing, and preparation of reports. Photography is desirable. The study is best accomplished by small groups of participants, each of which should include (or have access to) qualified botanists and reliable bird identifiers. Plots must be fairly accurate as to measurements, area, and topography, but amateurs are frequent participants. Reports must follow a standardized format and are published in the January issue of *American Birds.* An instruction reprint is available.

EARLY WINTER RANGE MAPPING

The Editor, *American Birds,* National Audubon Society, 950 Third Avenue, New York, N.Y. 10022

This program maps the distribution and abundance of species of birds found north of Mexico during the last two weeks in December. Selecting one or more recent years, range maps are constructed from the Christmas Bird Count data and published in *American Birds.* Each map is for a single species. Although this is desk work (instructions and materials are available), many volunteers have participated, which has resulted in a published book with 134 species maps.

NORTH AMERICAN HAWK MIGRATION STUDIES

Hawk Migration Association of North America, Box 51, Washington, Conn. 06793

These studies advance the knowledge of bird of prey migration across the continent. They provide, through the use of standard reporting forms and procedures, a bank of data on hawk migration for the use of professional and amateur ornithologists. They also help establish baselines for future monitoring of hawk populations. The Hawk Migration Association defends the right of birds of prey to living space and survival on the planet.

Observers record the numbers and species of hawks that migrate past known hawk migration lookouts, such as rooftops, mountain ridges, airports, beaches, and other places with a good view of the sky. Data are recorded on standard forms, which are postage paid and provided free. In areas where the lookout approach is not satisfactory, observers often perform road counts and similar surveys. Completed data sheets and other reports are sent to the appropriate regional editor. From this material, the editors write seasonal reports, which are published along with other hawk-watching information in a semiannual newsletter. The

145. The turkey vulture is one bird of prey observed by the North American Hawk Migration Studies.

field records are keypunched into computers at the Migratory Bird and Habitat Research Laboratory in Laurel, Maryland, and then are forwarded to Hawk Mountain Sanctuary in Kempton, Pennsylvania, where they are stored for future use by professional and amateur researchers.

Participants should be able to recognize hawks in flight and record field data. Observers are urged to be conservative in their identifications and careful in their counting and record keeping. Participants vary widely in experience, but even beginning hawk-watchers are welcome and encouraged to participate.

INTERNATIONAL SHOREBIRD SURVEY
Manomet Bird Observatory, Manomet, Mass. 02345, *or* Canadian Wildlife Service, 1725 Woodward Drive, Ottawa, Ontario K1G 3Z7

This survey counts shorebirds at a convenient location and at prescribed intervals. It compares routes and timing of migration for North American shorebirds, particularly those species east of the Rocky and Andes mountain ranges. The survey is conducted in more than fifteen New World countries from Canada to Argentina. The survey also helps to locate color-marked individuals and identify major stopover and migration staging areas. Survey results are used for conservation, research, and management purposes. Participants must have a good ability to identify and estimate numbers of shorebirds.

146. The International Shorebird Survey helps to identify major stopover and migration staging areas from Canada to Argentina. Photo by O. S. Pettingill/Cornell Laboratory of Ornithology

PACIFIC SHOREBIRD PROJECT
G. W. Kaiser, Canadian Wildlife Service, P.O. Box 340, Delta, British Columbia
   The Pacific Shorebird Project consists of searching western shorebird flocks for individuals with colored bands, tags, or dye marks in order to improve our understanding of the movements of shorebirds. More than five thousand shorebirds (dunlin, western sandpiper, and least sandpiper) have been color banded in Vancouver, British Columbia, without a single sighting in the United States or Mexico. Participants should be able to identify most shorebirds.

ALL-SEASON PROGRAMS

NORTH AMERICAN BIRD BANDING PROGRAM
Bird Banding Laboratory, U.S. Fish and Wildlife Service, Laurel, Md. 20811, *or*
Bird Banding Office, Canadian Wildlife Service, 2721 Highway 31, Ottawa, Ontario K1A 0H3

Bird banding is the technique of placing return-addressed, serially numbered leg bands on birds and maintaining banding and subsequent-encounter records for study purposes. The technique relies on people to report voluntarily subsequent encounters with banded birds. An encounter—which is simply an observation of a previously banded bird—may occur as one hunts a bird, finds a bird dead, traps one, or sights it. The North American Migratory Bird Banding Program is cooperatively but independently administered by the Canadian Wildlife Service and the U.S. Fish and Wildlife Service. The Bird Banding Laboratory, located at Laurel, Maryland, serves as the central depository of records arising from the North American Bird Banding Program.

The bird banding program depends on the cooperation of volunteers both to band birds and to report recovered birds. Participants mark birds by means of sequentially numbered bands for the purpose of studying migratory patterns and population dynamics of living birds. If you find a dead banded bird, remove the band, straighten it out, and tape it securely to a piece of heavy writing paper. Send in the following information with the band: your name and address (plainly printed); all numbers and letters on the band; the date you found the band; the place you found the band (including mileage and direction from the nearest town, with county and state); and how you found the band (on a dead bird, shot or caught in some other way). Place the band and information in an envelope, mark the envelope "Hand Cancel," and send to Bird Banding Laboratory, U.S. Fish and Wildlife Service, Laurel, Md. 20811. If you find a live banded bird, carefully read the number on the band, write it down, and release the bird. Mail the band number and the above information to the Bird Banding Laboratory.

You will receive a Certificate of Appreciation from the laboratory telling you where the bird was banded, what kind it was, and who banded it. The person who banded the bird will also learn where and when you found the bird.

Anyone at least eighteen years of age who knows how to identify all the common birds in their different seasonal plumages may apply for a banding permit from the U.S. Fish and Wildlife Service. The applicant must furnish the names of three well-known bird banders or ornithologists who can vouch for his or her fitness as a bird bander. Only persons well qualified and with plans for research are issued banding permits.

There are currently five bird banding associations in North America. Their function is to coordinate banding activities within their region and to provide training workshops to improve the banding skills of their members. These organizations also publish journals that feature information learned about birds through banding. The regional associations welcome serious amateurs who want to learn banding skills and assist active banding programs. To find the location of your nearest banding station, contact the regional bird banding association in your region (see addresses below) or the Bird Banding Laboratory or the Canadian Wildlife Service at the addresses given above.

Eastern Bird Banding Association, Davis Corkran, 154 Lake Avenue, Fair Haven, N.J. 07701. Copublisher of the bimonthly journal *Bird Bander* with Western Bird

147. Participants in the North American Bird Banding program mark birds with sequentially numbered bands for the purpose of studying migratory patterns and population dynamics of living birds.

Banding Association. This abundantly illustrated journal contains a variety of semitechnical papers presenting the results of bird banding research and techniques of interest to banders. It also contains sections on recent literature and reports from banding stations. Membership includes subscription.

Northeastern Bird Banding Association, Inc., Membership Secretary, Box 797, Manomet, Mass. 02345. Publisher of *Journal of Field Ornithology* (formerly *Bird-Banding*). As its new name suggest, this excellent professional journal includes a variety of bird research topics in addition to banding studies. The serious amateur will find many topics of interest in this quarterly journal, but should be prepared for a thorough, quantitative approach to many of the subjects. Membership includes subscription.

Inland Bird Banding Association, Bernard N. Brouchoud, P.O. Box 456, Manitowoc, Wis. 54220. Publisher of the bimonthly magazine *The Inland Bird Banding News*, which contains reports from regional banding stations and interesting results from banding research. Membership includes subscription.

Western Bird Banding Association, Otis D. Swisher, 1002 South Oakdale, Medford, Oreg. 97501. Copublisher of *Bird Bander* with Eastern Bird Banding Association (see above). Membership includes subscription.

Ontario Bird Banding Association, David Brewer, R.R. #1, Puslinch, Ontario N0B 2J0. Publisher of *Ontario Bird Banding*, a journal emphasizing topics of interest to banders that also publishes articles and short notes on most aspects of ornithology. Membership includes subscription.

BACKYARD WILDLIFE HABITAT PROGRAM
National Wildlife Federation, Washington, D.C. 20036

Participants in this program develop suitable wildlife habitats on parcels of property smaller than 3 acres. Ample food, water, and cover must be provided. If a participant's yard meets the qualifications, he or she will receive a certificate, a quarterly newsletter, a state-by-state listing of other program participants, and, if requested, free booklets developed by the program staff. The program is available to any property holder or occupant of North America. Information about local topics, such as exotic flora and fauna in Florida, local weed ordinances, prairie restoration in the Midwest, and landscaping with drought-resistant plants in the Southwest, is found in the quarterly newsletter.

ATLANTIC AND GULF COAST BEACHED BIRD SURVEY PROJECT
Malcolm M. Simons, Jr., 1701 E. Harbor View Road, Box 52, Charlotte Harbor, Fla. 33950

The project has participants who walk beaches to count, identify, and record numbers of dead birds. The data are used to establish a baseline of "normal" seabird mortality, as measured by dead birds on the beaches. The information will be used to measure changes—either short-term changes caused by natural or man-made disasters, or long-term changes caused by gradual changes in the environment. The survey area currently runs from New England down the East Coast to southern Florida and around the Gulf Coast to Padre Island, Texas. The project may be extended to the Atlantic coast of Canada. No degree of expertise is required, since dead birds cannot fly away, and if problems of identification arise, the specimens can always be taken to a more knowledgable observer or to a university or museum. Directions and report forms are available.

GREAT LAKES BEACHED BIRD SURVEY
Mr. Chris Risely, Long Point Bird Observatory, P.O. Box 160, Port Rowan, Ontario N0E 1M0, Canada

This survey monitors bird mortality in the Great Lakes region and obtains baseline environmental data to help in understanding the extent and causes of

148. *Volunteers in the Atlantic and Gulf Coast Beached Bird Survey Project walk beaches to count, identify, and record the number of dead birds.* Photo by W. Frederick Lucas

die-offs and normal mortality. Participants walk a 1- or 2-mile stretch of beach on the Great Lakes once a month, preferably for a year or more, and count and identify any dead or dying birds that are encountered. The information is record-ed on survey forms that are turned in every four months. Park personnel, game wardens, and others who patrol beaches regularly are particularly invited to take part in the survey.

WEST COAST BEACHED BIRD CENSUS

Point Reyes Bird Observatory, 4990 State Route 1, Stinson Beach, Calif. 94970

This census helps determine the seasonal distribution of aquatic birds in near-shore waters by using beach-cast carcasses as an index. It monitors mortality rates of birds in the same waters, particularly in relation to oil pollution. It censuses the same strip of West Coast beach once a month for at least one year. Partic-ipants should be able to identify aquatic birds in the hand.

AMERICAN ROBIN SURVEY

Rev. R. Charles Long, S. Cedar Drive, Apartment 409, Scarborough, Ontario
M1J 3E6, Canada

The American Robin Survey helps accumulate basic life history and behavior data on the American robin. Although the robin is abundant and much has been written about it, large gaps in the knowledge of the bird remain. Specific information, such as arrival and departure dates, nesting dates, number of broods, and latest date for young in nest is reported.

SEASONAL FIELD OBSERVATION REPORTS

The Editor, *American Birds,* National Audubon Society, 950 Third Avenue, New York, N.Y. 10022

The Seasonal Field Observation Reports record unusual bird occurrences during the spring and fall migrations in the United States and Canada. They provide data for a summary entitled "The Changing Seasons," which reviews the nesting season, winter season, and the migration from a continent-wide perspective with consideration of the effects of weather and winter survival. They report unusual bird sightings, low numbers, and high numbers to the regional editor. "The Changing Seasons" summary and the migration reports for each region are published in *American Birds.* To locate the appropriate regional editor, write to the address given above.

BLUE LIST

The Editor, *American Birds,* National Audubon Society, 950 Third Avenue, New York, N.Y. 10022

Participants determine which species (and subspecies) are experiencing population declines in all or part of their range. Such declines being apparently noncyclical and the species not yet having reached the status of "endangered," the Blue List is an early warning list. Participants' report forms are published each year in the May issue of *American Birds.* Observers with at least five years of experience in their reporting area nominate bird species that, in their opinions, belong on the Blue List. The official Blue List is published in the November issue of *American Birds.*

ANALYTICAL AND/OR DERIVATIVE STUDIES

The Editor, *American Birds,* National Audubon Society, 950 Third Avenue, New York, N.Y. 10022

Participants in these studies analyze and present conclusions based on the masses of raw distribution data published by *American Birds* in its Christmas Bird Count and seasonal regional field observation reports. Opportunities are limitless for deriving important studies from published Christmas Count data, which include studies of single species range expansions or contractions, incursions, population increases or decreases, data relating to species migrations, patterns in rare occurrences, vagrants (birds outside their normal range), and so on. Considerable data is available in computer data banks. Participants should have a reasonably advanced ornithological interest.

OPERATION BIRD (BIRDING IN RIPE DUMPS)
ED – Room 435 PSB, 231 W. Michigan, Milwaukee, Wis. 53201

Sanitary landfills (refuse dumps) are becoming important wildlife attractions as they proliferate in number and grow in size. They provide an abundant food supply, and where water bodies have been created and woodlands are interspersed, a diversity of habitats are created for a wide variety of birds. The attractiveness of dumps to birds has even become a hazard to human beings in certain locations, such as those near airports or municipal water supplies.

Participants in Operation Bird map the location of dumps throughout the upper Midwest. They describe the attributes that make them attractive to birds and formulate management plans to improve birding opportunities at dumps. They also document species presence and abundance by conducting regular censuses. Participants include professionals and amateurs from private industry, universities, and nature centers. A quarterly newsletter, *The Scavenger,* is sent to all participants; instructions and record sheets are also provided. Future plans include studies of sewage-treatment plants and associated ponds and lagoons, as well as computerization of data to facilitate long-term studies of population trends.

WATERFOWL HABITAT CONSERVATION
Ducks Unlimited, Inc., P.O. Box 66300, Chicago, Ill. 60666

There are more than twelve hundred Ducks Unlimited chapters in the United States and Canada. Members observe waterfowl migration, detailing dates, nesting activities, and staging (grouping before migration), and commenting on local habitat conditions (e.g., water availability) for waterfowl production. Participants may obtain landowners' viewpoints on waterfowl numbers and migration and habitat conditions in a local area. They also make recommendations for conserving wetlands. Ducks Unlimited's principal concern is conservation of the prairie provinces of Canada where most North American waterfowl breed. An effort is also directed toward preserving wetlands in Mexico, where much of North America's waterfowl population winters. Participants should have an interest in waterfowl, an ability to identify birds, and knowledge of the local area.

U.S. STATE AND LOCAL PROGRAMS

Arizona

SOUTHWEST WILDLIFE REHABILITATION PROJECT
Kathryn A. Ingram, DVM, 3133 E. Bloomfield, Phoenix, Ariz. 85032

This project treats orphaned and injured birds, most of which are banded and released (nonreleasables are used for education or research) and welcomes interested volunteers in the vicinity of Phoenix.

149. Ducks Unlimited, Inc., promotes waterfowl conservation through habitat preservation programs in the United States, Canada, and Mexico.

RAPTOR STUDY COMMITTEE
Arizona Wildlife Federation, P.O. Box 27573, Phoenix, Ariz. 85061

The committee's purpose is to discover amd monitor population trends of Arizona raptors by keeping nesting records and data on sightings for all species. Participants must have some expertise in identifying raptors, and there are some opportunities for participation in raptor banding.

SOUTHWEST HAWK WATCH
Dr. Sally Spofford, Southwest Hawk Watch Committee, P.O. Box 106, Portal, Ariz. 85632

Participants in the Hawk Watch study population trends of migrating raptors in south central and southeast Arizona. On prescribed spring and fall dates, they travel established routes by vehicle and/or maintain observation posts on certain mountaintops, where participants identify and count raptors and record weather data, observation conditions, and so on. Participants should have some expertise in identifying raptors or should accompany experienced observers. There are also occasional opportunities to assist in raptor banding.

150. *The Southwest Hawk Watch in Arizona studies population trends of migrating raptors.*

BIOLOGY OF BREEDING BALD EAGLES IN ARIZONA

Dr. Robert D. Ohmart, Department of Zoology, Arizona State University, Tempe, Ariz. 85281

Sponsored by state and federal agencies at Salt and Verde rivers in central Arizona, this program collects information on the biology of breeding bald eagles in Arizona by having participants sit all day at various stations to collect information on feeding and nesting behavior. More active observers can participate in radio-tracking the movements of instrumented young. Participants should be patient observers and capable of taking careful field notes.

Arkansas

ANNUAL WARBLER MIGRATION STUDY

Dr. Earl L. Hanebrink, Box 67, State University, Ark. 72467

Sponsored by the Northeast Arkansas Audubon Society, the study identifies and counts warblers and all other birds at Big Lake Wildlife Refuge in Manila, Arkansas. Participants conduct the count from mid-April to mid-May of each year, and the annual checklist of all species of birds compiled during the count is cataloged at the Arkansas State University Department of Biology. (This study was originally a search for Bachman's warbler.)

151. *A search for the rare Bachman's warbler initiated the Arkansas Annual Warbler Migration Study.*

California

AVIAN RESEARCH PROGRAM
Avian Science, University of California at Davis, Davis, Calif. 95616

This program conducts research into various bird topics, including studies of poultry, game birds, raptors, and exotic birds in the vicinity of Yolo and Solano counties. Many projects require assistance from interested volunteers.

SPRING BIRD COUNT
Davis Audubon Society, P.O. Box 886, Davis, Calif. 95616

This is a census of the spring bird population in Yolo and Solano counties. Participants record the behavior of nesting birds.

WILDLIFE REHABILITATION PROGRAM
Davis Audubon Society, P.O. Box 886, Davis, Calif. 95616

Also sponsored by the Davis Wildlife Care Association, participants in the program assist with the rehabilitation of injured birds and other wildlife in the Davis vicinity and help conduct educational programs for the public. No qualifications are necessary other than an interest in wildlife.

SAN ELIJO LAGOON MONTHLY BIRD SURVEY/CENSUS
John DeBeer or Mona Baumgartel, San Elijo Lagoon Bird Survey Group/San Diego Field Ornithologists, 1630 Burgundy Road, Leucadia, Calif. 92024

The survey/census records bird activity in this vulnerable ecological area. Participants identify and count birds on the first Sunday of each month, and beginners are welcome.

STUDIES IN AVIAN COMMUNITY ECOLOGY
Point Reyes Bird Observatory, 4990 State Route 1, Stinson Beach, Calif. 94970

This program includes long term studies of land bird communities in coastal northern California, particularly Point Reyes National Seashore and its vicinity. Participants conduct censuses and observe bird behavior. Opportunities to assist bird banders are available, and participants must be able to identify birds and record exact information. Living facilities are available for long-term volunteers or weekend workers.

Colorado

COLORADO WINTER/SPRING POPULATION COUNTS
Chuck Loeffler, Colorado Division of Wildlife, 2126 N. Weber, Colorado Springs, Colo. 80907

This is a long-term study of the spring and winter bird populations of the Colorado Springs Wildlife Management Area near Fountain, Colorado, conducted by the Division of Wildlife and the Aiken Audubon Society. Participants conduct counts similar to the Audubon Christmas Bird Count to determine species abundance and habitat preference in the cottonwood-riverbottom and cactus-grassland

152. The Colorado Bird Distribution Latilong Study collects bird distribution and nesting data for developing nongame management plans.

habitats. Group leaders and assistants are needed for both the January count and the May count.

COLORADO BIRD DISTRIBUTION LATILONG [LATITUDE/LONGITUDE] STUDY
Nongame Section, Colorado Division of Wildlife, 6060 Broadway, Denver, Colo. 80216

Sponsored by the Colorado Division of Wildlife and the Colorado Field Ornithologists, this program collects bird distribution data for the development of nongame management plans and environmental assessments throughout Colorado. Participants collect data on bird distribution and nesting occurrences. Amateur participants with an ability to identify birds are the backbone of this program. Data are stored on a computer for three to four years and then retrieved for evaluation and summary.

Connecticut

BIRD BANDING
Milan G. Bull, Connecticut Audubon Society, 314 Unquowa Road, Fairfield, Conn. 06430

Persons of college age with a strong interest in a wildlife-related field are welcome to assist a permanent bird banding station in southern Connecticut (Fairfield County) in the maintenance of bird trapping sites, recording data, and handling birds. Experience is not necessary.

LEAST TERN NESTING SURVEY

Milan G. Bull, Connecticut Audubon Society, 314 Unquowa Road, Fairfield, Conn. 06430

College students in related fields are welcome to assist in maintaining fencing and information signs for the protection of a small nesting colony of least terns (*Sterna albifrons*) and in the collection of breeding data. Participants keep a beachside vigil outside the colony and educate the public. Arrangements can be made for academic credit.

WOOD DUCK NESTING SURVEY

Milan G. Bull, Connecticut Audubon Society, 314 Unquowa Road, Fairfield, Conn. 06430

This is a survey and management program for the wood duck on several Audubon sanctuaries in Connecticut (Fairfield and Montville counties), aimed at identifying the breeding population and sanctuary usage by this species. Participants help to erect and maintain nest boxes, update yearly records and maps, and identify nesting habitats.

Delaware

BLUEBIRD NESTBOX PROJECT

Mr. Bill Zimmerman, 815 W. 22nd St., Wilmington, Del. 19802

The project provides nest boxes and checks nesting success of bluebirds in southern sections of Pennsylvania and nothern Delaware (New Castle Co.). Participants build boxes, place them in appropriate habitats, and census them each week. After the nesting season, participants clean and repair boxes and compile census cards.

LEAST TERN PROJECT

Mrs. Marguerite Jahn, 603 Cambridge Dr., Newark, Del. 19711

Participants walk beaches to help protect least terns. Volunteers are needed to patrol tern nesting areas and to assist with educational efforts. The work requires dedicated participants who can spend many long, hot hours censusing and protecting the terns.

TRI-STATE BIRD RESCUE

Lynne Frink, Coordinator, P.O. Box 1713, Wilmington, Del. 19899

Sponsored by the Delaware Audubon Society, this program helps to rescue and rehabilitate oil-damaged waterfowl on both sides of the lower Delaware River and Upper Delaware Bay (southeast Pennsylvania, south New Jersey, and Delaware). It specializes in one or more of the following areas: field retrieval, cleaning, medical treatment, rehabilitation, and operations control. Participants must be over eighteen and attend a Tri-State Bird Rescue training session in one of the specialized areas. The Rescue has more than two hundred trained volunteers and is the officially recognized "rescue" organization on the lower Delaware River.

*153. The Florida Tern and Skim-
mer Nesting Colony Survey protects
tern and skimmer colonies from vandalism and human disturbance.*

Florida

TERN AND SKIMMER NESTING COLONY SURVEY

Dr. Herbert W. Kale II, Florida Audubon Society, 35-1st Court S.W., Vero
Beach, Fla. 32960

This survey locates all tern and skimmer colonies in Florida, determines the
number of breeding pairs, and protects each colony from vandalism and human
disturbance. Participants record data on the number of birds, ownership of land
on which the colony is located, and the exact location of the colony. Anyone who
can recognize terns and skimmers may assist, and qualified individuals may conduct more detailed observations of the breeding biology and population dynamics
of each colony. To avoid unnecessary disturbances, do not enter colonies unless
advised by the society.

COLONIAL WADING BIRD SURVEY

Dr. Herbert W. Kale II, Florida Audubon Society, 35-1st Court S.W., Vero
Beach, Fla. 32960

This survey locates wading bird colonies throughout Florida to determine species composition and population densities. Participants determine the exact location of each colony and ownership of the site and record the nesting substrate and
surrounding habitat. Participants must be able to identify various species of herons, ibis, cormorants, and pelicans. More detailed observations may be desirable,
depending on accessibility of the colony and qualifications of the collaborator. To
avoid unnecessary disturbances, do not enter colonies unless advised by the society.

Georgia

SONGBIRD MANAGEMENT AREA CENSUS

Harriet G. Di Gioia, U.S. Forest Service, 401 Old Ellijoy Road, Chatsworth, Ga. 30705

Volunteers in the census assist with a study of the effects of timbering on birds and other wildlife in Chattahoochee National Forest in northwest Georgia. A two-year pretimbering census has been made, and after-timbering censuses are now conducted.

CHATTAHOOCHEE HAWK WATCH

Harriet G. Di Gioia, U.S. Forest Service, 401 Old Ellijoy Road, Chatsworth, Ga. 30705

The Hawk Watch counts hawks during the fall migration over the Cohutta Mountains, part of Georgia's Blue Ridge Mountains in the northwest part of the state. Participants assist with documentation of Georgia hawk migration patterns and should be able to recognize hawks in flight or be willing to assist in spotting birds.

SPRING BIRD COUNT

Harriet G. Di Gioia, U.S. Forest Service, 401 Old Ellijoy Road, Chatsworth, Ga. 30705

Conducted with the Cherokee Audubon Society, the Spring Bird Count is a day of bird counting to determine population trends of breeding birds in Whitfield County and the Songbird Management Area of Chattahoochee National Forest (Murray County). Beginners are welcome, but bird identification skills are helpful.

BALD EAGLE WATCH

Harriet G. Di Gioia, U.S. Forest Service, 401 Old Ellijoy Road, Chatsworth, Ga. 30705

This is a census of the winter population of bald eagles at Carter's Lake and other habitats in northwest Georgia conducted by the U.S. Forest Service and other agencies. Participants view films about eagles and are given directions detailing where they can observe eagles in Georgia. No special qualifications are needed.

Hawaii

ENDANGERED SPECIES PROGRAM

Dr. C. John Ralph, Institute of Pacific Islands Forestry, 1151 Punchbowl Street, Honolulu, Hawaii 96813

This program of the institute and the U.S. Forest Service determines the ecological role of native and exotic forest birds to help manage endangered species in Hawaii statewide and in the Hawaii Volcanoes National Park. Volunteers participate in a research program for periods of three months or longer that involves daily work in the field recording data on behavior, feeding habits, and breeding activities. Participants must be in good physical condition and should expect fairly rough conditions with daily rain in the forest. A bachelor's degree in biology or equivalent training or experience is necessary. Dormitory housing can be arranged for volunteers during most periods.

WINTER AND SUMMER WATERBIRD SURVEYS
Mr. Ronald L. Walker, Wildlife Branch, Hawaii Division of Fish and Game, 1151 Punchbowl Street, Honolulu, Hawaii 96813

These surveys determine the abundance and distribution of waterbird species throughout Hawaii. Participants must be able to identify migrant and resident waterbirds found in Hawaii. Surveys are conducted twice a year—in January and July.

Idaho

BANDING FISH-EATING BIRDS IN IDAHO
Dr. C. H. Trost, Biology Department, Idaho State University, Pocatello, Idaho 83209

Volunteers in the program band as many herons and ibis as possible to locate their wintering grounds and determine the cause or causes of high mortality. Participants must be in good physical condition and willing to wade in mosquito-infested marshes to catch young herons for banding during late May and June.

Illinois

EAGLE VALLEY HAWK WATCH
Terrence N. Ingram, Eagle Valley Environmentalists, Box 155, Apple River, Ill. 61001

During the third weekend in September, volunteers count and identify species of hawks migrating over Eagle Valley, a major midwestern flyway for migrating hawks. Observers have good opportunities to photograph migrating hawks. In some years, hawks are trapped, banded, and released. Beginners are welcome.

BALD EAGLE COUNT AT EAGLE VALLEY
Terrence N. Ingram, Eagle Valley Environmentalists, Box 155, Apple River, Ill. 61001

Volunteers in the project count wintering bald eagles on the Mississippi River at and near Eagle Valley and the power plants in Cassville, Wisconsin, and document bald eagle feeding patterns and use of the habitat. They also determine the number of eagles using Eagle Valley as a roosting area. No special qualifications are required.

NORTH CENTRAL ILLINOIS BIRD DISTRIBUTION STUDY
North Central Illinois Ornithological Society, 813 N. Main Street, Rockford, Ill. 61103

This study documents the abundance and distribution of birds in north central Illinois for the eventual publication of *Birding in North Central Illinois.*

SUMMER BIRD COUNT

North Central Illinois Ornithological Society, 813 N. Main Street, Rockford, Ill. 61103

Sponsored by the society and the Southern Illinois Bird Observatory, this count is a statewide effort to document the breeding distribution of Illinois birds. Beginners are welcome to join experienced observers.

ANNUAL STATEWIDE SPRING BIRD COUNT

Vernon M. Kleen, Division of Wildlife Resources, Illinois Department of Conservation, 100½ E. Washington Street, Springfield, Ill. 62701

Participants in this annual spring census of bird species and population distributions in Illinois spend all or any portion of the day (midnight to midnight) in the field identifying and tabulating individual birds and species seen. Participants should have some knowledge of local birds. The count is held on the first Saturday of May.

PREPARATION OF BIRD CHECKLISTS FOR STATE-OWNED PROPERTIES

Vernon M. Kleen, Division of Wildlife Resources, Illinois Department of Conservation, 100½ E. Washington Street, Springfield, Ill. 62701

This program determines the distribution of birds on state-owned properties so that management plans will conserve important bird habitats. Participants must identify the location and habitats of breeding species, especially the less common ones.

MID-JUNE BIRDING CHALLENGE

Vernon M. Kleen, Division of Wildlife Resources, Illinois Department of Conservation, 100½ E. Washington Street, Springfield, Ill. 62701

Participants in the challenge assemble detailed information about the abundance and distribution of bird populations in Illinois and help determine the presence and/or breeding status of birds (usually within close proximity to the observer's home) in order to determine management decisions for specific areas.

NEST RECORD CARD PROGRAM

Vernon M. Kleen, Division of Wildlife Resources, Illinois Department of Conservation, 100½ E. Washington Street, Springfield, Ill. 62701

Participants in the program record detailed information about nesting Illinois birds in cooperation with Cornell University's Nest Record Card Program, including information about nests on special record cards.

FIELD NOTE CONTRIBUTIONS

Vernon M. Kleen, Division of Wildlife Resources, Illinois Department of Conservation, 100½ E. Washington Street, Springfield, Ill. 62701

Participants record all information on bird abundance and distribution through

Illinois and must be able to identify birds, record observations, and submit field notes. Field notes are summarized for each season, and contributors are credited in the account for their observations.

BLUEBIRD TRAIL
Mrs. Gill Moreland, McHenry County Audubon Society, P.O. Box 67, Woodstock, Ill. 60098

This program provides nest boxes for eastern bluebirds, monitors nesting success, gathers data on site preference and predation, and bands young. No special qualifications are necessary.

Iowa

WINTER RAPTOR SURVEY
Dean M. Roosa, Preserves Advisory Board, Wallace State Office Building, Des Moines, Iowa 50319

The Preserves Advisory Board gathers information on wintering Iowa raptors to provide population estimates, trend indicators, and habitat requirements. These data will be used to establish management recommendations to protect Iowa raptors. Participants record activity, habitat, age, and number of raptors observed within one township in one day. Surveys are run between mid-January and mid-February on a still, clear, moderate day, preferably by two or more persons. It is open to anyone with an interest in observing, and beginners are paired with experienced observers.

IOWA LATILONG [LATITUDE/LONGITUDE] WILDLIFE SURVEY
Nongame Wildlife Biologist, Wildlife Research Section, Iowa Conservation Commission, Wallace State Office Building, Des Moines, Iowa 50319

This survey assesses the current status, distribution, abundance, habitat associations, and requirements of nongame wildlife in Iowa. It also monitors population changes and provides the information gathered to managers, decision makers, environmental assessors, and the public. Participants report observations of uncommon wildlife including such information as species, location, habitat, sex, and age. It is open to anyone with the ability to identify Iowa birds. Write for wildlife observation forms.

FREMONT COUNTY BIRD FORAY
W. Ross Silcock, Route 2, Malvern, Iowa 51551

This is an annual survey of early June breeding bird populations in Fremont County in the southwestern corner of Iowa. Participants count breeding birds in forests, fields, and marshes, and survey bird populations on foot, by car, by canoe, and while wading. Some competence in bird identification is desirable, but anyone with sharp eyes or a willingness to record data is welcome.

Kentucky

KENTUCKY ONE-DAY BALD EAGLE COUNT
Mrs. Anne Stamm, 9101 Spokane Way, Louisville, Ky. 40222

Volunteers count bald and golden eagles throughout Kentucky to estimate the winter population and search an area of prospective eagle habitats and report results and weather conditions. To avoid unnecessary disturbances, be certain to coordinate assistance with the contact.

Maine

COMMON LOON SURVEY
Maine Audubon Society, 118 Route One, Falmouth, Maine 04105

This survey provides data about the status of the common loon in Maine and educates the users of Maine's lakes on how they may coexist with loons. Participants can assist by conducting surveys on lakes for breeding loons.

LEAST TERN PROTECTION PROGRAM
Maine Audubon Society, 118 Route One, Falmouth, Maine 04105

This program attempts to protect the least tern colonies on the sandy beaches of southern Maine. Participants help construct fencing and other barriers around nesting sites to protect terns and piping plovers from dogs and human interference. They also monitor the tern area during peak public use of beaches.

154. Participants in the Maine Least Tern Protection Program construct fencing and other barriers around nesting sites.

MAINE BREEDING BIRD ATLAS
Maine Audubon Society, 118 Route One, Falmouth, Maine 04105

The atlas documents the breeding distribution of bird populations of Maine. Participants catalog nesting and breeding records for birds in specifically assigned areas within the state. Ability to identify birds is necessary.

BALD EAGLE NESTING AND WINTERING SURVEY
School of Forest Resources, University of Maine, Orono, Maine 04469

Also sponsored by the U.S. Fish and Wildlife Service and the Maine Department of Inland Fisheries and Wildlife, this survey locates all bald eagle breeding sites and winter concentration areas in Maine.

Maryland

NORTH POINT HAWK WATCH
Maryland Ornithological Society, Baltimore Chapter, Clyburn Park Mansion, 4915 Greenspring Avenue, Baltimore, Md. 21209

The watch determines the extent of the hawk migration over Baltimore County by monitoring the migration at North Point and Fort Howard.

DAN'S ROCK HAWK WATCH
T. Paul Smith, 909 Camden Avenue, Cumberland, Md. 21502

Sponsored by the Allegany Chapter of the Maryland Ornithological Society, the watch counts migratory birds of prey during the fall and spring migrations. Participants should be able to recognize birds of prey in flight.

MARYLAND BLUEBIRD NEST BOX PROJECT
Dr. Lawrence Zeleny, 4312 Van Buren Street, Hyattsville, Md. 20782

Sponsored by the Maryland Ornithological Society, the project constructs, provides, and monitors bluebird nest boxes. Participants submit an annual report of nesting activity.

MARYLAND BREEDING BIRD ATLAS
Chandler S. Robbins, 7900 Brooklyn Bridge Road, Laurel, Md. 20810

Sponsored by the Maryland Ornithological Society, the atlas maps the local breeding distribution of Maryland birds. Participants must attend workshops, become familiar with habitats in an assigned grid block, and observe and record breeding birds in specified areas.

BLUEBIRD NEST BOX PROJECT
Kendrick Y. Hodgdon, 945 Weires Avenue, La Vale, Md. 21502

Sponsored by the Allegany Chapter of the Maryland Ornithological Society, the project provides a trail of bluebird nest boxes at the Carey Run Sanctuary in Eckhart, Maryland. Participants can assist with construction, repair, and maintenance of nest boxes as well as monitor nesting success. No special qualifications are needed.

155. The Massachusetts Nighthawk Migration Watch records nighthawks passing by in the evening.

Massachusetts

ASHBY BIRD BANDING PROGRAM

Peter M. Johnson, Camp Middlesex, 65 Erickson Road, Ashby, Md. 04131

Conducted by the Ashby Bird Observatory, participants in the program assist professional bird banders in the collection of data about bird migration and other bird research topics. Housing is available year round. Beginners are welcome to assist this program, located in rural north central Massachusetts.

NIGHTHAWK MIGRATION WATCH

Massachusetts Audubon Society, Lincoln, Mass. 01773

Participants in the watch record the size and extent of the nighthawk migration in Massachusetts by observing nighthawks passing by a backyard or other site, generally between 7:00 and 8:30 in the evening. Weather and direction of flight are also recorded.

MASSACHUSETTS BREEDING BIRD ATLAS

Massachusetts Audubon Society, Lincoln, Mass. 01773

The atlas maps the breeding distribution of all species of birds nesting in Massachusetts. Participants were assigned one or more blocks of roughly 10 square miles. During the atlas period, each species was assigned one of three categories—possible, probable, or confirmed breeder. Many participants had no special qualifications other than the ability to identify birds. The original project period was 1974–1979, and may be repeated at a future date.

Michigan

MICHIGAN HERON COLONY SURVEY

Alice H. Kelly, 3681 Forest Hill Drive, Bloomfield Hills, Mich. 48013

Participants in the survey collect information on the location of active and historic heron colonies in Michigan, reporting locations (including county, township, distance to nearest road, plus your name, address, and telephone number) to the contact.

DETROIT AUDUBON SOCIETY BIRD SURVEY

Mrs. Neil T. Kelley, Chairman, Bird Survey Committee, 3681 Forest Hill Drive, Bloomfield Hills, Mich. 48013

Sponsored by the Detroit Audubon Society, the survey is a continuing study of migration dates, abundance, and status of birds and has been in existence since 1945 within selected counties of Michigan and Ontario. Participants report all bird sightings on prepared forms, and summaries are sent to *American Birds* and the state ornithological journal, *Jack-Pine Warbler.*

DETROIT AUDUBON SOCIETY NEST CARD PROGRAM

3681 Forest Hill Drive, Bloomfield Hills, Mich. 48013

This program is a file of nesting data in cooperation with the Cornell Nest Card Program. Continuous since 1945, it finds and checks nests without disturbing the birds and fills out cards with significant data (more than one visit to a nest is preferable). Participants should have experience at record keeping and the ability to identify birds.

BALD MOUNTAIN COUNT

Ellie T. Cox, Oakland Audubon Society, 18310 Sunderland, Detroit, Mich. 48219

The count documents the distribution and status of birds at Bald Mountain Recreation Area near Lake Orion in Oakland County. Members of the count participate in the annual count day in late May and also submit bird observations throughout the year. Participants should have a good ability to identify local birds and keep accurate records of birds (and other fauna). Beginners are welcome to improve their skills by accompanying experienced observers.

HOLIDAY BEACH HAWK WATCH

Ellie T. Cox, 18310 Sunderland, Detroit, Mich. 48219

Sponsored by the Oakland Audubon Society and the Detroit Audubon Society, the watch gathers fall hawk migration data at Holiday Beach Provincial Park in Essex County near Amhertsburg, Ontario. Participants count and identify hawks during the daylight hours of September, October, and November, and occasionally in December. Participants must have the ability to identify hawk species and make accurate number estimates and submit observation data on prescribed forms. Beginners are welcome to accompany veteran observers as apprentice and/or record keeper. Spectacular flights of broad-winged hawks occur in mid-September. Buteos migrate through in October and November. Golden and bald eagles, Cooper's hawks, and goshawks are also observed.

MICHIGAN AUDUBON SOCIETY BIRD SURVEY

Doug McWhirter, Michigan State University, Department of Zoology, East Lansing, Mich. 48824

The survey compiles reports of observations made throughout Michigan to document bird species present, their abundance, and their migratory movements. Observers should use standardized Michigan Audubon Society report forms and return them at specified times. In addition to sightings, report rare observations, such as effects of weather and habitat change on bird populations.

MICHIGAN BIRD FEEDER SURVEY
Michigan Audubon Society, 7000 N. Westnedge Avenue, Kalamazoo, Mich. 49007

Participants in this survey identify and count all birds taking food or water in the immediate vicinity of their backyard bird-feeding areas one day each month from November to April. Participants need not count every month to participate. Write the society for registration and report forms.

BARN OWL CENSUS
Michigan Department of Natural Resources, Wildlife Division, Box 30028, Lansing, Mich. 48909

Once fairly common in Michigan, the barn owl is now listed as a threatened species. Report observations of numbers, nesting status, and location of sighting, including your name and address with reports.

PROJECT LOON WATCH
Dr. Herbert Lenon, 4370 W. Remus Road, Mount Pleasant, Mich. 48858

Sponsored by the Chippewa Valley Audubon Club, participants document the distribution and abundance of breeding loons in Michigan by watching for loons on freshwater lakes during the summer breeding season.

INDEPENDENCE OAKS BIRD SURVEY
Kathleen Dougherty, 2800 Watkins Lake Road, Pontiac, Mich. 48054

Sponsored by the Oakland Audubon Society, participants document the distribution and status of birds occurring in Independence Oaks County Park in northwest Oakland County. They also take part in the annual count day—the first Sunday in June. Beginning bird enthusiasts may improve their field skills by accompanying experienced observers.

SOUTHFIELD BREEDING BIRD CENSUS
Betty Challis, Oakland Audubon Society, 20340 Westpointe Court, Southfield, Mich. 48076

Participants in the census gather data on the status of breeding bird populations at Southfield Civic Center Nature Area in South Oakland County. They accompany compilers on early morning censusing trips and indicate on a map of the census area the location of singing males. Participants must have the ability to identify songs of most breeding birds.

Minnesota

MINNESOTA BIRD FEEDER SURVEY
Doug and Julie Keran, Route 7, Box 14, Brainerd, Minn. 56401

Participants in the survey identify and count feeder birds every other weekend from October through April to monitor winter bird movements throughout Minnesota and provide estimates of bird populations. Participants should be able to identify common feeder birds. Write for an information packet and count procedures.

BLUEBIRD RECOVERY PROJECT

Audubon Chapter of Minneapolis, P.O. Box 566, Minneapolis, Minn. 55440

The purpose of the project is to increase the bluebird population in Hennedin, Sherburne, Wright, and Anoka counties. Participants assist with the construction of new nest boxes, monitor nesting success (five visits per year), and evaluate progress of the project.

MINNESOTA BIRD DISTRIBUTION REPORTS

Seasonal Report Editor, *The Loon,* James Ford Bell Museum of Natural History, 16 Church Street, S.W., University of Minnesota, Minneapolis, Minn. 55455

Sponsored by the Minnesota Ornithologist's Union, participants collect information on the distribution, abundance, and migration of birds in Minnesota, which is published quarterly in *The Loon.* Send reports to the Seasonal Report Editor.

HAWK RIDGE RAPTOR WATCH

Minnesota Ornithologist's Union, Bell Museum of Natural History, 10 Church Street, S.E., Minneapolis, Minn. 55455

Sponsored by the Minnesota Ornithologist's Union and the Duluth Audubon Society, the watch monitors the large fall hawk migration over Duluth. It also observes hawk banding. Beginners can learn hawk identification skills and receive background information on raptor biology at special lectures.

NONGAME WILDLIFE PROGRAM

Nongame Wildlife Supervisor, Minnesota Department of Natural Resources, 300 Centennial Building, 658 Cedar Street, St. Paul, Minn. 55155

The program's purpose is to improve nongame wildlife habitats and collect data on the distribution and numbers throughout Minnesota. Volunteers can assist several ongoing projects, including a loon survey and a colonial waterbird census.

GREATER SANDHILL CRANE SURVEY

Minnesota Department of Natural Resources, 300 Centennial Building, 658 Cedar Street, St. Paul, Minn. 55155

The survey assesses the current status and distribution of the greater sandhill crane in Minnesota so that appropriate research and management efforts can be applied. Data are collected for both migratory and breeding cranes. Participants complete a special field data card, supplied by the Department of Natural Resources, for all observations of cranes. These cards are submitted by observers on the first day of May, August, and December.

UNCOMMON WILDLIFE REPORT PROGRAM

Nongame Wildlife Supervisor, Minnesota Department of Natural Resources, 300 Centennial Building, 658 Cedar Street, St. Paul, Minn. 55155

156. The Minnesota Greater Sand-hill Crane Survey collects data for both migratory and breeding cranes.

The program collects information on the abundance and distribution of uncommon wildlife in Minnesota. Participants record observations on standard field cards provided by the Department of Natural Resources. A list of uncommon wildlife species is available on request.

Missouri

BLUEBIRD MONITORING PROGRAM
Missouri Department of Conservation, P.O. Box 180, Jefferson City, Mo. 65102

The program compiles information on the abundance, nesting behavior, and reproductive success of eastern bluebirds in Missouri. It also compares nest box designs and placements, notes nest box use, and, where possible, records date of laying, clutch size, date of hatching, and number fledged for each box. Participants must have the ability to monitor a series of nest boxes carefully, many of whom have already established bluebird trails.

Montana

MONTANA HERON AND CORMORANT SURVEY AND CENSUS
Larry S. Thompson, 117 Pine, Helena, Mont. 59601

This program is a census of all breeding colonies of herons and cormorants in Montana. Participants visit colonies during the breeding season to describe exact locations and count active nests. To avoid unnecessary disturbances and to coordinate visitation at colonies, write to the address above.

157. The New Hampshire Wood Duck Nest Box Program builds and maintains nesting boxes in areas where sufficient natural nesting cavities are not present.

New Hampshire

WOOD DUCK NEST BOX PROGRAM
Harold C. Lacaillade, New Hampshire Fish and Game Department, 34 Bridge Street, Concord, N.H. 03301

This program helps increase the wood duck population in areas where sufficient natural nesting cavities are not present. Participants construct, erect, and maintain wood duck nest boxes, and can also assist by transporting boxes to favorable wetland areas and erecting them on trees and stubs. Basic carpentry skills and tools are necessary.

BALD EAGLE MIDWINTER SURVEY
Audubon Society of New Hampshire, 3 Silk Farm Road, Concord, N.H. 03301

The survey takes a census of wintering bald and golden eagles in New Hampshire and assists in a state waterfowl tally. Participants note bald and golden eagles, separating as to age, if possible. The route followed should be described, noting latitude and longitude, landmarks, time, and weather. Waterfowl and other species of interest should also be noted. Beginning participants should accompany experienced observers.

NEW HAMPSHIRE HAWK WATCH
Carol F. Smith, Audubon Society of New Hampshire, 3 Silk Farm Road, Concord, N.H. 03301

Participants in the watch trace routes and determine numbers of hawks. Beginners are encouraged to work with experienced hawk watchers before conducting their own counts.

LOON PRESERVATION PROJECT
Scott Sutcliffe, Loon Preservation Committee, Humiston Building, Main Street, Meredith, N.H. 03253

Volunteers assist professionals in determining loon nesting success, survival of the young, and other biological information. No special qualifications are needed.

New Jersey

BREEDING HERONS IN SUSSEX COUNTY
Thomas McLaughlin, Raccoon Ridge Bird Observatory, Box 81, Layton, N.J. 07851

Volunteers record the size of breeding heron populations over a ten-year period at two heronries (Lafayette and Montague) in Sussex County in northwest New Jersey. Participants may assist with observation and photography of breeding herons in the two colonies.

BIRD BANDING PROGRAM
Dorothy W. Huges, Raccoon Ridge Bird Observatory, Box 81, Layton, N.J. 07851

Volunteers assist with the trapping and netting of birds at the observatory. No special qualifications are needed, but volunteers should purchase recommended books and plan on spending long hours in the field.

BREEDING HAWK CENSUS IN SUSSEX COUNTY
Mrs. Laura Socha, Raccoon Ridge Bird Observatory, Box 81, Layton, N.J. 07851

Volunteers record the breeding hawk population over a ten-year period in Sussex County in northwest New Jersey. Participants must learn the breeding habits of hawks and where they reside in the county and assist with banding hawks and recording data.

MONTCLAIR HAWK WATCH
Andrew Bihun, Jr., 18 Normal Avenue, Upper Montclair, N.J. 07043

Sponsored by the Montclair Bird Club, the watch counts and identifies the hawks that migrate past the Montclair lookout, located 15 miles west-northwest of New York City. The count is conducted from September 1 to November 30, and participants maintain a daily watch from nine o'clock in the morning to five o'clock in the evening. The only qualifications are a good pair of binoculars, a folding chair, and patience. Beginners are assisted by experienced observers.

New Mexico

SPRING ARRIVAL PROGRAM
J. R. Travis, 9410 Avenida de la Luna N.E., Albuquerque, N. Mex. 87111

Sponsored by the New Mexico Ornithological Society, the program determines spring arrivals of selected species in New Mexico. Operating since the late 1960s, the program's data have been contributed mostly by amateurs.

ROADSIDE RAPTOR SURVEY
J. P. Hubbard, 2016 Valle Road, Santa Fe, N. Mex. 87501

Sponsored by the New Mexico Game and Fish Department, the survey conducts roadside counts of raptor birds to determine population trends, distribution, and other aspects of their status. Volunteers fill out forms on road trips, giving mileage by county for each date. Observations should include species, age, sex, color phases, and number for all birds of prey. Participants must be able to identify raptors.

BREEDING BIRD PROGRAM
J. P. Hubbard, 2016 Valle Road, Santa Fe, N. Mex. 87501

Sponsored by the New Mexico Ornithological Society, the program documents detailed breeding behavior of birds in New Mexico. Participants must be able to identify breeding birds and complete breeding record cards. The data collected are made available to the Cornell Nest Record Card Program.

158. The New Mexico Roadside Raptor Survey conducts roadside counts to determine raptor population trends and distribution.

New York

ALLEGANY COUNTY BREEDING BIRD SURVEY
Clarence Klingensmith, R.D. 1, Alfred Station, N.Y. 14803

The Allegany County Bird Club counts birds along one or more 12-mile routes. Each of the eighteen routes is surveyed once during June to study relationships between bird populations and environmental changes. Participants must be able to recognize by sight and sound the birds that normally breed in western New York State.

GREAT SOUTH BAY NESTING BOX PROGRAM
Jack Foehrenbach, Great South Bay Audubon Society, Box 37, Brightwaters, N.Y. 11718

The program provides nesting sites for cavity-nesting birds in the towns of Islip and Babylon. Participants put up new houses, clean and rehabilitate existing houses, and monitor and record nesting success. Assistance is required primarily in spring and early summer.

ENDANGERED SPECIES PROGRAM
Endangered Species Unit, New York State Department of Environmental Conservation, Wildlife Resources Center, Delmar, N.Y. 12054

The purpose of this program is to clarify the status of listed endangered or threatened migratory birds, such as bald eagles, peregrine falcons, and ospreys. Participants help determine the population status, the movements of certain marked individuals, and the present breeding status of potentially endangered or threatened species in New York State, such as the northern harrier, Cooper's hawk, golden eagle, and roseate tern. Program volunteers monitor nesting success in cooperation with department personnel and determine the locations and size of nesting populations of certain colonial nesting birds. Participants must be able to identify birds as well as make detailed observations. Write for copies of the current observation form and details about how you can assist this program.

SIGNIFICANT HABITATS PROGRAM
Significant Habitat Unit, New York State Department of Environmental Conservation, Wildlife Resources Center, Delmar, N.Y. 12054

This program protects significant fish and wildlife habitats by public acquisition, legal protection through regulation, and by the environmental review process. Volunteers describe significant habitats on special report forms to nominate areas for protection. Write for copies of the appropriate form and further information on how to make nominations.

NEW YORK STATE AVIAN RECORD COMMITTEE
Laboratory of Ornithology, Cornell University, 159 Sapsucker Woods Road, Ithaca, N.Y. 14850

In conjunction with the Federation of New York State Bird Clubs, the committee documents the occurrence of rare or unusual species of birds in New York

State. Participants must be familiar with identification of New York State birds and reporting procedures of the committee. An annual report is published in *The Kingbird.* Write for reporting forms and descriptions of procedures.

NEW YORK STATE BREEDING BIRDS ATLAS PROJECT
Dr. C. R. Smith, Laboratory of Ornithology, Cornell University, 159 Sapsucker Woods Road, Ithaca, N.Y. 14850

Also sponsored by the Federation of New York State Bird Clubs and the New York State Department of Environmental Conservation, this project documents the breeding distribution of New York State birds. Volunteers search selected 6-square-mile blocks to determine the breeding status of resident birds. They also help identify for preservation critical habitats with especially unusual or rich assemblages of breeding birds. Participants should have the ability to identify by both sight and sound most of the state's breeding birds or assist experienced observers.

NEW YORK BIGHT PELAGIC BIRD STUDY PROJECT
P.O. Box 19992, New York, N.Y. 10008

This project studies the factors affecting the distribution and seasonal abundance of pelagic birds in the study area (offshore New York and New Jersey, Hudson Canyon, and Shark Gut Shoals). Participants record data on ocean and weather variables and act as observers on cruises. They must be seaworthy and available for extended periods (two weeks), and will be asked to sign a waiver of responsibility. Single individuals are preferred, and observer training and seasickness pills are provided.

HOOK MOUNTAIN HAWK WATCH
Stiles Thomas, 201-327-4600 or 201-327-3470

The Watch assists the Society for the Preservation of Birds of Prey count migratory hawks that pass Hook Mountain in Nyack, N.Y. Volunteers count the hawks with the assistance of trained observers and enjoy the splendid scenery of the Hudson Highlands.

159. Observers in the New York Bight Pelagic Bird Study Project check the distribution and seasonal abundance of birds in offshore New York and New Jersey waters.

ANNUAL SPRING MIGRATION SURVEY

Dr. Gordon M. Meade, 27 Mill Valley Road, Pittsford, N.Y. 14534

Sponsored by the Federation of New York State Bird Clubs, this program surveys migrant and resident birds statewide during a specified period in May. It identifies and counts species seen in an area of the participant's choice. Participants must be familiar with bird identification and counting procedures. The annual report appears in *The Kingbird.*

DUTCHESS COUNTY NEST BOX PROGRAM

Mr. and Mrs. Jones W. Key, 52 Tree Tops Lane, Poughkeepsie, N.Y. 12603

Sponsored by the Ralph T. Waterman Bird Club, volunteers help construct, maintain, and monitor eastern bluebird nest boxes. No special qualifications are necessary.

CAYUGA COUNTY BLUEBIRD TRAIL

Michael P. Riley, R.D. 1, Skaneateles, N.Y. 13152

Sponsored by the Owasco Valley Audubon Society, volunteers provide nest boxes for bluebirds in southern Cayuga County. Participants build, clean, and erect houses and census nesting success.

ANNUAL MIDWINTER WATERFOWL COUNT

Walton B. Sabin, 652 Kenwood Avenue, Slingerlands, N.Y. 12150

Sponsored by the Federation of New York State Bird Clubs, participants identify and count waterfowl occurring within New York State during a specified period in midwinter and submit a report of their findings. The count usually occurs during a specified period in January. The annual report appears in *The Kingbird.*

FIELD OBSERVATIONS OF NEW YORK STATE BIRDS

Emanuel Levine, Editor, *The Kingbird,* 585 Mead Terrace, South Hempstead, N.Y. 11550

Sponsored by the Federation of New York State Bird Clubs, participants document the distribution and abundance of the birds of New York State. They submit their reports with documentation to the appropriate regional editors of *The Kingbird.*

North Carolina

AMERICAN KESTREL POPULATION CENSUS

Dr. Richard D. Brown, Department of Biology, University of North Carolina at Charlotte, Charlotte, N.C. 28223

This is a census of winter and breeding American kestrels in the south central piedmont of North Carolina (Anson, Cabarrus, Iredell, Mecklenburg, Rowan, Stanly, and Union counties). Volunteers aid in the census of this species, which has "special concern" status in North Carolina. Participants complete field note forms; the data aid in the ongoing study of movements, habitat preference, and other life history considerations. Write for field note sheets.

AMERICAN KESTREL AND EASTERN BLUEBIRD NEST BOX PROGRAM
Dr. Richard D. Brown, Department of Biology, University of North Carolina at
Charlotte, Charlotte, N.C. 28223

Volunteers help in the production, placement, maintenance, and monitoring of
kestrel and bluebird nest boxes in the south central piedmont region of North
Carolina to help increase breeding populations of these declining species.

North Dakota

BIRDS OF BURLEIGH COUNTY
Genevieve Buresh, Bismarck-Mandan Bird Club, 1527 N. 19th Street, Bismarck,
N. Dak. 58501

This program determines the distribution of breeding and migratory birds in
Burleigh County. Volunteers observe and report birds, noting specific townships
of occurrence. Participants must be able to identify all birds encountered and to
locate the position of observation on county maps.

Ohio

OHIO RAPTOR SURVEY PROJECT
Ohio Biological Survey, 484 W. 12th Avenue, Columbus, Ohio 43210

This survey amasses scientifically defensible baseline data on Ohio's raptors,
assesses the impact of various land-use patterns on raptors, and determines raptor
productivity in Ohio. Participants complete specially designed observation cards
and/or nest cards describing the habitat where raptors are observed. Anyone with
the ability to identify Ohio raptors and provide accurate data may participate.

Oklahoma

AMERICAN WOODCOCK SURVEY
Dr. John S. Barclay, Department of Ecology, Fisheries, and Wildlife, 417 Life
Sciences West, Oklahoma State University, Stillwater, Okla. 74074

This survey monitors the changes, status, and distribution of the woodcock on
a seasonal and annual basis within Oklahoma. Participants report all sightings of
woodcocks, their nests, and their broods. Reports should include the name and
address of the person who made the observation, location of the sighting (county,
range, township, and section), date sighted, number of birds involved, and general
description of the habitat. Spring sightings should also include mention of breeding displays if observed. Hunter harvest records are welcome. Observers must be
able to distinguish the woodcock from similar species.

SPRING BIRD COUNT
Helen C. Miller, School of Biological Science, Oklahoma State University,
Stillwater, Okla. 74074

160. *The Oklahoma American Woodcock Survey monitors the status of the woodcock on a seasonal and annual basis.*

Sponsored by the Payne County Audubon Society, the count monitors local and migrant bird populations in Payne County. Participants count species and numbers of individuals following techniques similar to the Audubon Christmas Count. Anyone with an interest in birds is welcome to participate and learn from experienced leaders.

Oregon

RECORDS OF OREGON BIRDS
Southern Willamette Ornithological Club, Box 3082, Eugene, Oreg. 97403
This is a collection of bird distribution data for Oregon that is summarized for publication in *Oregon Birds* and will be used to compile an Oregon State checklist. Report all bird observations of interest to the address above.

EUGENE BLUEBIRD TRAIL
Eugene Natural History Society, Box 3082, Eugene, Oreg. 97403
Volunteers improve the bluebird population near Eugene by constructing and maintaining a trail of bluebird nest boxes within a 15-mile radius of the city. No special qualifications are necessary.

Pennsylvania

BLUEBIRD CONSERVATION AND EDUCATION PROJECT
Albert Thomas, 50 Alexander Drive, R.D. #2, Irwin, Pa. 15642
Sponsored by the North American Bluebird Society, volunteers construct bluebird houses and special sparrow houses to control sparrow competition. Volunteers also maintain records concerning the occupancy of the bluebird houses and assist qualified banders with bluebird banding. Slide shows are presented to school, church, and senior citizen groups. Anyone with an interest in bluebird

conservation is welcome to assist this program located in northeastern Pennsylvania.

HAWK MIGRATION STUDIES

Hawk Mountain Sanctuary Association, Route 2, Kempton, Pa. 19529

This program monitors hawk populations and gathers information about numbers and species that migrate past the Hawk Mountain Sanctuary. Participants take notes on the kinds of hawks, their number, and their behavior. Skills in hawk identification and note taking are necessary.

AMERICAN KESTREL NESTING STUDY

Hawk Mountain Sanctuary Association, Route 2, Kempton, Pa. 19529

This study involves field observation and banding of American kestrels at nest sites near the Hawk Mountain Sanctuary. Participants make sunrise-to-sunset observations of nesting behavior. Patience and an interest in observing the details of nesting behavior are necessary.

WARREN COUNTY BLUEBIRD NEST BOX PROJECT

William L. Highhouse, 8 Fourth Avenue, Warren, Pa. 16365

The purpose of this project is to increase the number of eastern bluebirds nesting in Warren County by placing nest boxes throughout the county. Participants purchase or construct a number of nest boxes, which are placed in the proper habitat and monitored throughout the nesting season. Twelve thousand bluebirds have fledged from boxes in the twenty-two-year history of this project.

South Carolina

STATE BREEDING BIRD SURVEYS OF SELECTED SPECIES

Nongame and Endangered Species Section, South Carolina Wildlife and Marine Resources Department, P.O. Box 167, Dutch Plaza Building D, Columbia, S.C. 29202

This program gathers population data on species of special concern in South Carolina, such as the Mississippi kite, Wilson's plover, ground dove, and Swainson's warbler. Participants monitor selected populations of certain species and record field data. They must be able to identify birds in the field and record accurate field notes.

South Dakota

BLUEBIRD BOX

South Dakota Department of Wildlife, Parks, and Forestry, Anderson Building, Pierre, S. Dak. 57501

This project provides nest boxes for eastern and mountain bluebirds in South Dakota. Send a stamped, self-addressed envelope for nest box plans and instructions.

161. The Pheasant Restoration Program in South Dakota attempts to increase the statewide ring-necked pheasant population by improving dense nesting cover.

PHEASANT RESTORATION PROGRAM

South Dakota Department of Wildlife, Parks, and Forestry, Anderson Building, Pierre, S. Dak. 57501

This program's purpose is to increase the statewide population of the ring-necked pheasant by improving dense nesting cover, restocking in areas devoid of wild birds, and occasional control of problem predators. Participants enroll 10 to 40 acres of land for planting of dense nesting cover and must agree not to disturb this cover for three years. The participant must own land in South Dakota; monetary remuneration is provided (approximately $22 per acre).

Texas

BLACKBIRD POPULATION STUDIES

Keith A. Arnold, Department of Wildlife and Fisheries Science, Texas A & M University, College Station, Tex. 77843

In conjunction with the Texas Agricultural Experiment Station, this program studies long-term population trends in wintering populations of cowbirds, grackles, and starlings in Central Brazos Valley and environmental manipulation as a means of resolving bird-human problems. Participants can aid in censusing blackbirds and starlings at feeding sites, in flight lines, and at roosts. They must be willing to spend odd hours in the field, and a brief training session is offered.

BIRDS OF THE CENTRAL BRAZOS VALLEY AND VICINITY

Keith A. Arnold, Department of Wildlife and Fisheries Science, Texas A & M University, College Station, Tex. 77843

The department and the Texas Cooperative Wildlife Collections at the university sponsor this program in which volunteers report sightings of birds from the

162. The Texas Blackbird Population Studies report on the wintering population trends of cowbirds, grackles, and starlings.

ten county areas surrounding the Central Brazos Valley. Participants must be familiar with birds of the areas, and reports are subject to rigorous scrutiny.

BIRD NESTING HABITS WITHIN FINITE BOUNDARIES
Rod Rylander, Red River Refuge, 800 N. Travis, Sherman, Tex. 75090

This program of the refuge (east of Denison) locates, identifies, and records the nesting activities of resident birds. Participants should have the ability to identify most nesting species on the preserve and observe movements to establish territories and locate nests.

Vermont

VERMONT BREEDING BIRD ATLAS PROJECT
Vermont Institute of Natural Science, Woodstock, Vt. 05091

This project's purpose is to describe the distribution of all birds breeding in Vermont and provide baseline ecological data to evaluate future land development in the state. Volunteers identify specific species and/or habitats that warrant intensive studies or protection. The Vermont Atlas Survey is based on mapping units of 10-square-mile blocks. Each area is surveyed for breeding species of birds that are classified as possible, probable, or confirmed nesters, depending on strict criteria of field evidence. All birders, including residents and summer visitors, are welcome to participate.

RECORDS OF VERMONT BIRDS
Vermont Institute of Natural Science, Woodstock, Vt. 05091

This program records and publishes seasonal bird observations from Vermont. Participants report species and date for all unusual bird sightings. Notable reports appear in the quarterly publication *Records of Vermont Birds.*

BANDING VERMONT BIRDS
Vermont Institute of Natural Science, Woodstock, Vt. 05091

Volunteers in this program assist in ongoing studies of bird migration patterns near south Woodstock. Research is concerned with such topics as longevity of native birds and eye color parameters of the white-throated sparrow and catbird. The banding program contributes data to *Records of Vermont Birds* and *American Birds.* Banding demonstrations are provided to visiting school groups, and full training courses are given to volunteers. Banding experience is helpful, but not necessary.

VERMONT HAWK WATCH
Vermont Institute of Natural Science, Woodstock, Vt. 05091

Also sponsored by the Vermont Audubon chapters, this program monitors the spring and fall hawk migration through Vermont, determines major routes, if any, and records the influence of weather and geography. Participants monitor predetermined sites on selected spring and fall dates. Beginners are welcome to assist experienced observers.

CARDINAL, TUFTED TITMOUSE, AND MOCKINGBIRD COUNT
Vermont Institute of Natural Science, Woodstock, Vt. 05091

This program determines the wintering populations of the cardinal, tufted titmouse, and mockingbird in Vermont. Participants report the largest number of each species at bird feeding stations on the second weekend of February.

Virginia

CHRISTMAS COUNT IN JUNE
Myriam P. Moore, 101 Columbia Avenue, Lynchburg, Va. 24503

Sponsored by the Lynchburg Bird Club and the Virginia Society of Ornithology Local Chapters Committee, this census is a June count of the Virginia Christmas Bird Count study areas. Participants should be able to identify and count most Virginia birds, or they can accompany an experienced observer.

BIRDS OF WILLIAMSBURG CHECKLIST
Bill Williams, 157 W. Queens Drive, Williamsburg, Va. 23185

Sponsored by the Williamsburg Bird Club, participants document the bird species found in the Williamsburg, James City County, and York County vicinity for the preparation of a checklist. They should be able to identify local birds, since full documentation is required for rare species.

PURPLE MARTIN POPULATION STUDY
Sam Hart, Ferncliff Drive, Williamsburg, Va. 23185

This is a study of the local population biology of the purple martin. Participants may assist with banding and cleaning nest boxes, and no special qualifications are needed.

BLUEBIRD NEST BOX STUDY
Bill Williams, 157 W. Queens Drive, Williamsburg, Va. 23185
  This study promotes the use of bluebird nest boxes to aid in the population expansion of the eastern bluebird. Participants assist in banding, nest box construction, and maintenance. No special qualifications are needed.

COLONIAL WATERBIRD BREEDING SURVEY
Bill Williams, 157 W. Queens Drive, Williamsburg, Va. 23185
  Sponsored by Nature Conservancy, this survey locates and monitors all colonial waterbird colonies on Virginia's barrier islands. Participants assist population surveys by visiting colonies during specified times. They must be able to identify colonial waterbirds and have an interest in colonial waterbird ecology. This study involves extensive volunteer field work.

HAWK MIGRATION STUDY
Bill Williams, 157 W. Queens Drive, Williamsburg, Va. 23185
  This study documents hawk flight patterns along the southern tip of Virginia's eastern shore peninsula and identifies and counts hawks on a daily basis from September 15 through early November. Participants must be able to identify raptors in flight.

Washington

WASHINGTON MIDWINTER BALD EAGLE SURVEY
Coordinator, Washington Department of Game, 600 N. Capital Way, Olympia, Wash. 98504
  This survey helps determine the distribution and relative population size of wintering bald eagles in Washington. Participants assist by counting eagles during a two-week period in one or more survey areas. They must have the ability to distinguish immature bald eagles from golden eagles.

WASHINGTON NATURAL HERITAGE DATA SYSTEM
Nongame Program-Data System, Washington Department of Game, 600 N. Capital Way, Olympia, Wash. 98540
  This program contributes information about vertebrate and invertebrate animals in Washington. Of special interest are 250 species that have been given "species of special interest" status. Write for a list of the species, instructions, and observation cards.

West Virginia

AUTUMN HAWK WATCH
Bibbee Nature Club, Box 125, Lerona, W. Va. 25971
  This watch determines which bird of prey species pass the East River Mountain near Bluefield, West Virginia, and records species and approximate numbers. Ob-

servers spend time on the mountain and report their findings to the Bibbee Nature Club. Participants must be able to identify hawks and take weather readings, and instruction is available. The watch is usually conducted from mid- to late September.

CENTURY DAY BIRD COUNT
Bibbee Nature Club, Box 125, Lerona, W. Va. 25971

The purpose of this count is to find as many species as possible in one day in the Pipestem area of West Virginia and to aid in the preparation of a checklist for this area. Volunteers can join the club's groups, and all bird reports are welcome. Instruction is offered to beginning bird-watchers.

BLUEBIRD NESTING BOX PROJECT
Bibbee Nature Club, Box 125, Lerona, W. Va. 25971

Participants in this project construct and erect bluebird houses and collect data on the use of the boxes in southern West Virginia.

Wisconsin

SEASONAL FIELD NOTES
Daryl D. Tessen, *Passenger Pigeon,* 2 Pioneer Park Place, Elgin, Ill. 60120

Sponsored by the Wisconsin Society for Ornithology, this program determines the abundance and distribution of all bird species in Wisconsin, and records appear in the state magazine *Passenger Pigeon* and *American Birds.* Participants record field sightings, including species, number, location, and documentation, the latter when the sighting involves an "exceptional" species or an unusually early or late date. Observers must have bird identification skills.

ENDANGERED AND THREATENED SPECIES OBSERVATIONS
Office of Endangered and Nongame Species, Wisconsin Department of Natural Resources, P.O. Box 7921, Madison, Wis. 53707

This program compiles information on the status and distribution of endangered and threatened species of birds in Wisconsin. Participants record observations and nest records of such species, including date, specific location, and activity of the observed species. Report observations to the address given above and write for a current list of Wisconsin endangered and threatened species.

Wyoming

WYOMING AVIAN ATLAS
Nongame Bird Biologist, Wyoming Game and Fish Department, Lander Office, 260 Buena Vista, Lander, Wyo. 82520

This program compiles known information on the distribution of birds in Wyoming. Participants contribute information on standardized forms summarizing their observations. Observers must be able to identify Wyoming birds and collect accurate field notes.

JACKSON HOLE BIRD LIST
Nongame Bird Biologist, Wyoming Game and Fish Department, 260 Buena Vista, Lander, Wyo. 82520

This program updates the species list and seasonal abundance of birds in the Jackson Hole–Grand Teton area. Participants record species observed on a checklist available at state and federal agencies in the Jackson area. Checklists, with additions, should be returned postage paid so that current observations can be added to the list.

CANADIAN PROVINCIAL AND LOCAL PROGRAMS

Alberta

MAY DAY BIRD COUNT
Federation of Alberta Naturalists, P.O. Box 1472, Edmonton, Alberta

This count documents the occurrence and distribution of birds throughout Alberta during the nesting season. It is conducted in the same manner as the Audubon Christmas Bird Count. During a predesignated weekend, generally the last in May, parties and/or individuals survey particular areas or routes, recording birds on standard forms provided by the coordinator. Inexperienced bird-watchers are generally teamed with more experienced observers. The results are published annually in the *Alberta Naturalist.*

LETHBRIDGE BLUEBIRD PROJECT
D. T. Mackintosh, 1919 9th Avenue S., Lethbridge, Alberta

Sponsored by the Lethbridge Naturalists Society and the Federation of Alberta Naturalists, this project encourages bluebirds to nest in the Lethbridge area. Volunteers build, locate, and maintain bluebird houses and monitor bluebird nesting success. It is open to anyone with an interest in bluebird conservation.

GOOSE HABITAT PROJECT
Michael O'Shea, 616 20th Street S., Lethbridge, Alberta T1J 3J2

Sponsored by the Lethbridge Naturalists Society and the Alberta government, this project encourages Canada geese and prairie falcons to nest on the cliffs in the vicinity of Lethbridge. Participants dig 2-foot-by-3-foot holes in cliffs for geese and smaller holes for falcons. They keep regular checks on holes for progress. A knowledge of roping is required, since some holes are located 250 feet above the Oldman River.

163. The Goose Habitat Project in Alberta provides new nest sites for Canada geese and prairie falcons on the cliffs in the vicinity of Lethbridge.

British Columbia

BLACK BRANT SURVEY

Mrs. J. Conway, 2147 S. Island Highway, Campbell River, British Columbia V92 2S9

Sponsored by the Mitlenatch Field Naturalist Society, this survey is a count from March to early May of the black brant on its spring migration past the east coast of Vancouver Island. No special qualifications are needed other than the ability to distinguish between adult and juvenile brants.

BIRDS IN REAL DISTRESS (B.I.R.D.)

Canadian Wildlife Service, P.O. Box 340, Delta, British Columbia U4K 3Y3

B.I.R.D. is a group of skilled and organized people who participate in oil-spill cleanups and bird rehabilitation on the lower mainland of British Columbia. Participants attend an S.P.C.A. training program and an oil-spill seminar, must be over age twelve and in good health, and must share a dedication to bird conservation.

PACIFIC SHOREBIRD BANDING

G. W. Kaiser, Canadian Wildlife Service, P.O. Box 340, Delta, British Columbia U4K 3Y3

This program determines the species and condition of migrating shorebirds in British Columbia and locates wintering and breeding grounds. Participants locate shorebird concentrations and assist experienced volunteers in banding, trapping, and records maintenance and can develop bird handling skills to "expert" status. They also assist with the trapping of birds using mist nets and canon nets, and must have the ability to withstand the boredom of long winter nights on the mud flats and be keen to learn and participate in a team project.

SHOREBIRD COUNT

G. W. Kaiser, Canadian Wildlife Service, P.O. Box 340, Delta, British Columbia U4K 3Y3

This count determines the movement of various species of shorebirds through British Columbia and Yukon Territory during the spring and fall migrations. Participants record numbers and species of birds on well-defined study areas. They must have the ability to identify common shorebird species and to estimate numbers accurately.

RAPTOR COUNT

Jude Grass, Vancouver Natural History Society, P.O. Box 3021, Vancouver, British Columbia U6B 3X5

This program monitors the raptor population, recording species changes, winter survival, and so on in the lower mainland of British Columbia. Participants conduct biweekly surveys of specific study areas and must have good identification skills and the ability to keep detailed and accurate field notes.

ATLAS OF BRITISH COLUMBIA BIRDS

R. Wayne Campbell and Michael G. Shepard, British Columbia Provincial Museum, Victoria, British Columbia V8V 1X4

This program determines the distribution (breeding and occurrence) of British Columbia birds. Participants provide field notes and/or museum index cards of sightings of all birds seen throughout the year. They must be able to identify most of the common birds occurring in British Columbia and provide detailed field records. All observations are closely examined and then added to a major data bank and retrieval system.

PHOTO-RECORDS FILE FOR BRITISH COLUMBIA VERTEBRATE RECORDS

R. Wayne Campbell, British Columbia Provincial Museum, Victoria, British Columbia V8V 1X4

This program documents by photos and film (black-and-white and color prints, 35mm slides, or movie film) the occurrence of rare birds, mammals, reptiles, and amphibians occurring in British Columbia. It obtains photographic evidence (not necessarily award-winning photos) of the occurrence of the animal with complete details concerning the record, date, number, location, photo equipment used, witnesses, and so on. At present, the file contains nearly seven hundred photos of rare vertebrates that have occurred in British Columbia over the past decade. Historical material is being added slowly. The file also documents the occurrence of three new Canadian species, at least twenty-five birds new to British Columbia, and many regional and local range extensions.

164. The Prairie Bluebird Nestbox Project has established a 1,600-mile trail of nest boxes in an attempt to increase bluebirds' numbers in the prairies of Manitoba and Saskatchewan.

Manitoba

PRAIRIE BLUEBIRD NESTBOX PROJECT
Mrs. John Lane, 1701 Lorne Avenue, Brandon, Manitoba R7A 0W2

Sponsored by the North American Bluebird Association and Friends of the Bluebird, this project extends from southwestern Manitoba into Saskatchewan and as far as Broadview and Churchbridge. It is an effort to bring the bluebirds back to the prairies by establishing nest boxes. Participants maintain and monitor some of the five thousand nest boxes erected along this 1,600-mile trail and also assist with the construction of new nest boxes. The project is open to anyone interested in and dedicated to bluebird conservation. Youth groups such as the 4-H Clubs and retired citizens are especially welcome.

Ontario

UNUSUAL BREEDING BIRDS OF ST. WILLIAMS REFORESTATION AREAS
Robert Curry, 92 Hostein Drive, Ancaster, Ontario L96 257

Sponsored by the Long Point Bird Observatory, this program helps compile data on the breeding status of birds nesting in this reforested sand plain region of Norfolk County. Participants must have a good knowledge of bird identification.

BIRDS OF ALGONQUIN PROVINCIAL PARK
Ron Tozer, Spring Lake Road, R.R. #1, Dwight, Ontario P0A 1H0

This program compiles an annotated list of the birds of the park, including information on past and present distribution and the status of each species. Participants send in past and/or recent observations from the park and take part in field work, such as the Christmas Count, Spring Round-up, and travel to interior areas.

BIRDS OF MOOSONEE CHECKLIST PROJECT

Alan Wormington, General Delivery, Gilford, Ontario L0L 1R0

Participants in this project help compile a detailed annotated checklist of the birds of the Moosonee/Moose Factory area and the adjacent James Bay flats east to the Quebec border and north to Cockispenny Point. They should submit observations, including numbers and locations for all species, and the compiler will gladly supply information to anyone planning a visit to the area on transportation, lodging, and so on.

BIRD LIST JAMES–HUDSON BAY LOWLANDS OF ONTARIO

District Manager, Ministry of Natural Resources, Box 190, Moosonee, Ontario P0L 1Y0

The program increases the knowledge of bird species composition, abundance, distribution, and status in the James–Hudson Bay Lowlands in Ontario, from the Quebec border to the Manitoba border. Visitors with a knowledge of bird identification should keep careful records of bird species, numbers, habitat, location, and behavior while in the area.

ONTARIO HERONRY INVENTORY

Long Point Bird Observatory, P.O. Box 160, Port Rowan, Ontario N0E 1M0

Also sponsored by the Canadian Wildlife Service and the Ontario Ministry of Natural Resources, the program is compiling a permanent catalog of all known heronries in Ontario. The catalog will provide a means of identifying key sites with a view to ensuring necessary protection. Population estimates will provide a baseline for measuring future population changes. Volunteers should report locations of past and present heronries. They will receive forms requesting information on the size of the colony, its history, habitat, ownership of the site, and present or potential human impact. Participants must have the ability to identify herons and record observations accurately. Discretion is essential to ensure minimum disturbance to heronries.

ONTARIO BIRD FEEDER SURVEY

Long Point Bird Observatory, P.O. Box 160, Port Rowan, Ontario N0E 1M0

The survey monitors bird populations at winter feeders throughout Ontario from October to April. Participants count birds at feeders, recording the largest number of each species seen at any one time during specified periods each month. Observers must be able to identify usual winter birds.

BANDING AND MIGRATION MONITORING

Warren Russell, Toronto Bird Observatory, 745 Gerrard Street E., Toronto, Ontario M5N 2L5

This program monitors the bird migration in the Toronto area. Participants assist the banding and censusing program in the spring (April to June) and fall (August to October). Both experienced banders and observers are welcome. Those without bird banding experience but with a serious interest in participating will receive training.

BIRD MORTALITY AT MANMADE STRUCTURES

David Broughton, 4 Heddington Avenue, Toronto, Ontario M5N 2L5

Sponsored by the Toronto Bird Observatory, this program monitors the mortality of migrants at man-made structures to determine why birds collide with certain structures. It helps to formulate methods for decreasing collisions. Participants visit selected sites during the migration seasons, collect dead birds, transfer them to a central location, and assist in the tabulation of results and preparation of specimens. All interested participants are welcome. Salvaged specimens are used for teaching and research.

LEO SMITH BLUEBIRD NESTBOX PROJECT

Leo S. Smith, 481 Waughan Road, Apartment 303, Toronto, Ontario M6C 2P6

This project maintains a bluebird nest box trail in Dufferin County, approximately 50 miles northwest of Toronto. Participants assist with the installation and maintenance of bluebird nest boxes in appropriate habitats.

Prince Edward Island

FRANCIS BAIN BIRDATHON

The President, Natural History Society of Prince Edward Island, P.O. Box 2346, Charlottetown, Prince Edward Island CIA 7N8

This program establishes basic information on breeding birds present on Prince Edward Island, including dates of arrival, species present, and so on. Participants run transects similar to those in breeding bird surveys, and also count birds away from the transects. Each team has at least one experienced birder, but experience is not required for everyone. The Birdathon is conducted on one day from sunrise to sunset during the last week in May.

FIELD CHECKLIST OF BIRDS, PRINCE EDWARD ISLAND

Parks and Conservation Branch, Department of Tourism, P.O. Box 2000, Charlottetown, Prince Edward Island C1A 7N8

Participants in this program establish species present, seasonal abundance, and breeding status and report sightings of birds, particularly new species or breeding records or any perceived change in the status for incorporation into the revised list. Sightings of common birds are accepted, usually without confirmation by other bird-watchers. New species and breeding records are usually scrutinized carefully.

Quebec

QUEBEC CHECKLIST PROJECT
Club des Ornithologues du Quebec, 8191 Avenue du Zoo, Charlesbourg,
Quebec G1C 4G4

Participants in this project report all bird observations in Quebec to maintain current distribution records. Records are entered into a computer and retrieved for various research programs, including updates of the Quebec checklist. Write for a copy of the current checklist.

Saskatchewan

INDIAN HEAD BLUEBIRD TRAIL
Lorne Scott, Box 995, Indian Head, Saskatchewan S0G 2K0

The purpose of this project is to increase the mountain bluebird and tree swallow populations by providing nest boxes. Participants study population biology and behavior through active banding programs and assist in recording nesting success and constructing nest boxes. No special qualifications are needed, and groups of young people are welcome to participate.

7

PERIODICALS AND ORGANIZATIONS

This chapter reviews the publications and activities of North American bird organizations and offers recommendations for selecting memberships. Ornithological organizations and their periodicals are presented in three groups—amateur bird-watcher periodicals with a broad North American focus, professional journals, and state and provincial periodicals.

NORTH AMERICAN PERIODICALS AND ORGANIZATIONS

Recommended

American Birds
Published six times a year by the **National Audubon Society.**

American Birds is dedicated to documenting the changing bird life of the North American continent. This includes the Hawaiian Archipelago, Bermuda, Mexico, Central America, the West Indies, and the northern rim of South America. As such, it is the principal journal in which both professionals and amateurs publish field observations about changes in distribution and unusual bird occurrences.

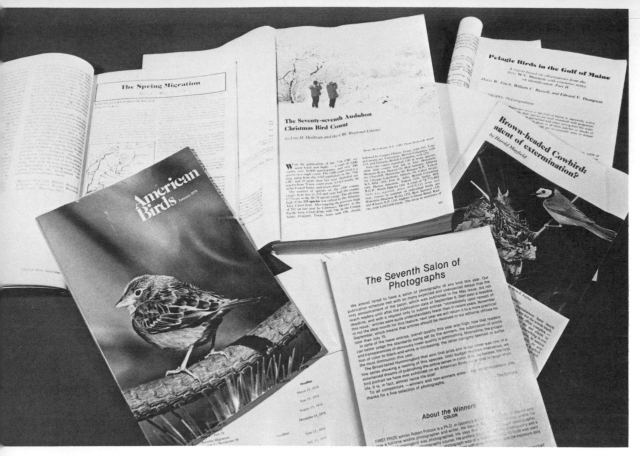

165. **American Birds** *is a recommended periodical published six times a year by the National Audubon Society.*

166. **The Living Bird** *is a recommended annual publication of the Cornell University Laboratory of Ornithology.*

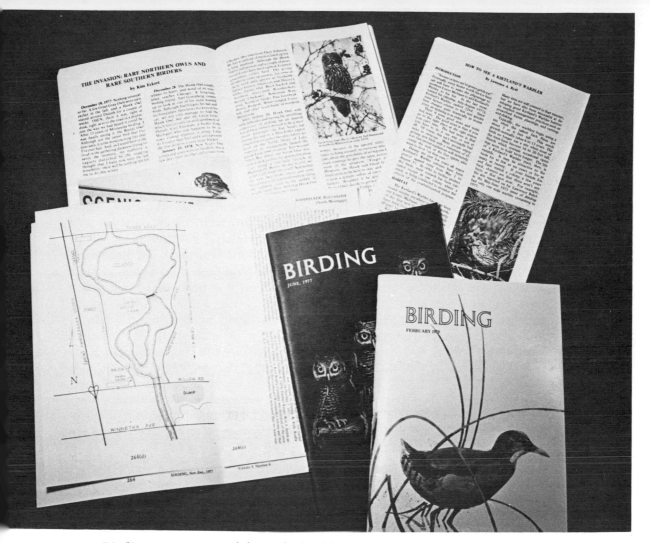

167. **Birding** *is a recommended periodical published six times a year by the American Birding Association.*

Each January issue reports the findings of the Winter Bird Population Study and the Breeding Bird Census (see pages 176 and 166). The March, May, September, and November issues contain observations respectively from the fall migration, winter season, spring migration, and nesting season. The July issue reports the results of the annual Christmas Bird Count (see page 174).

In addition to reporting changing patterns in bird numbers and distributions, *American Birds* also sponsors photo contests for bird photographers and contains such features as the column entitled "Going Places," which summarizes forthcoming bird tours. *American Birds* also publishes the annual Blue List, which serves as an early warning list of birds suspected of experiencing serious population declines. Most issues also contain field technique articles about such topics as identifying difficult species, selecting a telephoto lens, and taking field notes. For subscription contact: *American Birds,* 950 Third Avenue, New York, N.Y. 10022.

The Living Bird
Published annually by the **Cornell University Laboratory of Ornithology.** Members of the Laboratory of Ornithology receive *The Living Bird* and the quarterly *Newsletter to Members.*

The Living Bird offers a blend of broad-interest scientific research papers accompanied by art from the world's best-known bird artists. It features articles on a variety of bird biology topics, most of which will interest both serious amateurs and professionals. A sampling of titles includes "An Ecological and Behavioral Study of the Galapagos Penguin," "Biology of the Bald Eagle on Amchitka Island, Alaska," and "Breeding Strategies in Birds of Prey." Articles in *The Living Bird* have a minimum of statistics and quantitative discussion but an abundance of information that is well illustrated with black-and-white photographs, line drawings, and occasional full-page color illustrations. In addition to bird research papers, this journal also includes articles about people and their various interactions with birds. A few such titles are "Impact of Human Visitation on Avian Nesting Success," "With Louis Fuertes in Abyssinia," and "Recording Bird Song."

Membership in the laboratory also includes a subscription to *Newsletter to Members* and a quarterly update about ongoing educational and research programs of the laboratory. For membership or subscription contact: Cornell University Laboratory of Ornithology, 159 Sapsucker Woods Road, Ithaca, N.Y. 14850.

Birding
Published six times a year by the **American Birding Association, Inc.** Membership in the association includes subscription.

The American Birding Association exists to promote the hobby and sport of birding, and to educate the public toward a greater appreciation of birds. These objectives are achieved primarily through its publication, *Birding.*

Many of the articles in *Birding* concern the sport of listing birds. The magazine contains feature articles that detail where to find birds and how to identify difficult species. Each issue contains a section of inserts that can be removed and filed separately and that highlight especially good birding locales and detail travel directions to find unusual species. Each year the magazine publishes an annual supplement, which reports the world, state, and North American bird list totals of participating members.

In addition to articles about bird listing and finding birds, *Birding* contains reviews of new bird books and a column entitled "Gleanings from the Technical Literature." The magazine also publishes articles about bird-watching techniques. A few typical titles are "Answers to Your Questions on Binoculars," "A Budgeted Banquet for the Birds," and "Birding with a Questar."

The American Birding Association also publishes a checklist to North American birds and offers its members a discount on bird books. The association's sales department is an excellent place to purchase bird-finding publications and some regional checklists. Each year the association sponsors a national convention that features field trips to notable birding sites near the convention headquarters. For membership or subscription contact: American Birding Association, Inc., Box 4335, Austin, Tex. 78765.

Other Choices

Ornithological Newsletter

Published six times a year by the Ornithological Societies of North America (OSNA) for the American Ornithologists' Union, Cooper Ornithological Society, and Wilson Ornithological Society.

This one-page newsletter provides information on upcoming ornithological meetings, grants and awards, requests for research assistance, new publications, and other general announcements of interest to both professional and serious amateurs. It is available only to subscribers of either *The Auk, The Condor,* or *The Wilson Bulletin* (see pages 229–230). For subscription contact: The American Ornithologists' Union, c/o OSNA, 1735 Neil Ave., Columbus, Ohio 43210.

Birding News Survey

Published four times a year by Avian Publications, Inc.

Birding News Survey is the companion to *A Guide to North American Bird Clubs* (see page 260). This quarterly serves the useful function of reprinting articles and news from the multitude of local bird club journals and newsletters. These are organized under such headings as identification, field techniques, photography, habitat conservation, attracting birds, bird finding, and a calendar of selected ornithology events. Updated supplements to *A Guide to North American Bird Clubs* are frequently included in the form of removable inserts. For subscription contact: *Birding News Survey,* Avian Publications, Inc., P.O. Box 310, Elizabethtown, Ky. 42701.

Bird Watcher's Digest

Published six times a year by Pardson Corporation.

Printed on newsprint and styled after *Reader's Digest, Bird Watcher's Digest* is a compilation of selected newspaper articles about birds and the hobby of birdwatching. In the preface to the first edition, the publishers comment that in the process of selecting the thirty-two articles reprinted in their first edition, they reviewed a total of twenty-seven thousand articles from the abundant nontechnical newspaper media throughout North America. For subscription contact: *Bird Watcher's Digest,* P.O. Box 110, Marietta, Ohio 45750.

Continental Birdlife

Published six times a year by Continental Birdlife, Inc.

This journal is devoted to "the advancement and enjoyment of field ornithology." Its scope is all of Canada, Greenland, Mexico, and adjacent waters. Written for the serious amateur, *Continental Birdlife* features original articles about distribution studies, field identification techniques, important birding locations, and in-depth book reviews. A useful section entitled "Recent Literature" lists selected papers from the technical literature. For subscription contact: *Continental Birdlife,* P.O. Box 43294, Tucson, Ariz. 85733.

The Bird Watch

Published ten times a year by the *Bird Populations Institute.* Membership in the institute includes subscription.

The Bird Populations Institute publishes its four-page bulletin with the primary purpose of developing a deeper public understanding about wild bird populations. Written especially for amateur bird-watchers, the articles describe interesting features of bird life as they are interpreted by biologists. Sample titles include "Life at a Winter Blackbird Roost," "Feeding Strategy of the Dickcissel," and "How Birds Keep Warm at the Roost."

Bird Populations Institute offers a weekly question-and-answer column to newspapers about wild birds. It also provides a telephone number where callers receive advice about bird problems. For membership contact: Bird Populations Institute, P.O. Box 637, Manhattan, Kans. 66502.

Around the Bird Feeder

Published four times a year by the **Bird Feeders Society.** Membership in the society includes subscription.

Around the Bird Feeder is intended for the beginning bird-watcher who wants to know more about the common birds that visit backyard feeders. This pocket-sized quarterly averages twenty-five pages and contains a mixture of original articles and occasional reprints from other bird periodicals. Most articles contain tips on ways to attract particular species or solutions to feeder problems, such as squirrel invasions and moldy bird food. Each issue includes several letters from members and the editor's answers to bird feeding questions. Members receive a discount on selected bird feeding merchandise. For membership contact: Bird Feeders Society, P.O. Box 225, Mystic, Conn. 06355.

Wild Bird Guide

Published quarterly by **Bird Friends Society.** Membership in the society includes subscription.

Wild Bird Guide averages thirty pages and features articles about familiar birds. Several pages are devoted to reporting casual observations by members, and there is usually a section in which the editors respond to bird questions posed by readers. Each issue advertises discounted bird feeders, books, and records sold by the society. The narrow focus of the articles makes this a periodical exclusively for the backyard bird-watcher. Bird Friends Society also publishes leaflets about such topics as "How to Thwart Squirrels," "Cats and Other Pests," and "Recipes for the Birds." For membership contact: Bird Friends Society, Essex, Conn. 06426.

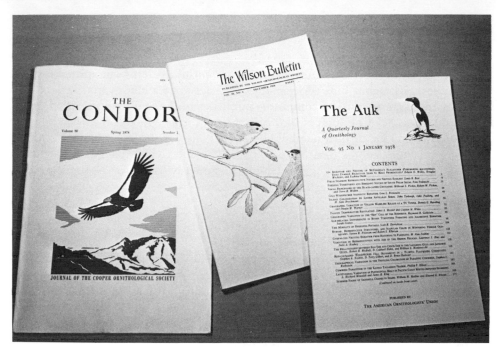

168. **The Condor, The Wilson Bulletin,** and **The Auk** are three professional ornithological journals that publish technical articles that will interest the serious amateur birder.

PROFESSIONAL JOURNALS

The Auk
Published four times a year by the **American Ornithologists' Union.**
Membership in the union includes subscription.

Each issue of *The Auk* contains approximately twenty scientific papers about various aspects of ornithology. The majority of them detail research concerning particular species as observed through field studies. The quantitative approach of most papers may discourage beginning birders, but any interested amateur can learn much from the concise abstracts that summarize each article.

In addition to the major papers, each issue also contains many brief notes titled "Short Communications." There the amateur can find interesting accounts of unusual bird behavior, such as "The Development of Shell-cracking Behavior in Herring Gulls," "Osprey Trapped by Water Chestnut," and "Male Pintails Defending Females from Rape." *The Auk* also includes a thorough book review section that includes comments on both popular and technical bird books. Each year the final issue is accompanied by a special supplement that reviews recent ornithological literature.

The American Ornithologists' Union also publishes in-depth ornithology studies in its publication series known as A.O.U. Ornithological Monographs. Another activity is the standardization of Latin and English bird names by the union's Committee on Classification and Nomenclature. This committee reviews current research about bird relationships and designates names that reflect similarities and differences. The committee publishes bird names and species descriptions in the A.O.U. Checklist of North American Birds, which is regarded by most as the "official" list of bird names. The A.O.U. also sponsors an annual national convention that features research paper sessions and field trips. For subscription contact: American Ornithologists' Union, National Museum of Natural History, Smithsonian Institution, Washington, D.C. 20560.

The Condor

Published four times a year by the **Cooper Ornithological Society.** Membership in the society includes subscription.

The Condor is a technical journal written primarily for professional ornithologists. The scientific papers concern all aspects of ornithology and include papers about birds from all parts of the world. In addition to the in-depth articles, there are shorter communications with the same detail as the full-length papers. Each issue also contains a section entitled "News and Notes," which notifies readers of upcoming ornithological meetings, and a recent publications section that reviews new bird books.

In addition to publishing *The Condor,* the Cooper Ornithological Society publishes *Studies in Avian Biology.* This is a series of longer, more in-depth papers that are published at irregular intervals. The society sponsors an annual meeting, usually held at an institution of higher learning in the western United States.

Amateurs may participate by attending annual meetings or by submitting scholarly articles for editorial review and possible publication in *The Condor.* Serious amateurs can also benefit by attending the annual meeting and exchanging information with the professional ornithologists who make up the bulk of the membership. For membership or subscription contact: Jane R. Durham, Treasurer, P.O. Box 520, Tempe, Ariz. 85281.

The Wilson Bulletin

Published four times a year by the **Wilson Ornithological Society.** Membership in the society includes subscription.

The Wilson Bulletin publishes scientific papers about the birds of the Western Hemisphere, with emphasis on North American birds. In addition to detailed studies on such topics as "Pair Formation Displays of the Great Blue Heron," "Molt in Leach's and Ashy Storm-petrels," and "The Effects of Orchard Pesticide Applications on Breeding Robins," each issue also presents about fifteen general notes, most of which describe unusual bird behavior.

Each issue also contains announcements of forthcoming ornithological meetings and a literature section that contains reviews of new books about birds.

The Wilson Ornithological Society, named after Alexander Wilson, the first American ornithologist, was founded in 1888 by youths interested in bird study. Today, its membership consists largely of professional ornithologists, but serious amateurs are welcome to join and participate by attending the annual meeting. *The*

Wilson Bulletin frequently accepts general notes written by serious amateurs who have taken the time to make careful field observations. For subscription contact: The Wilson Ornithological Society, c/o The Museum of Zoology, University of Michigan, Ann Arbor, Mich. 48104.

STATE AND PROVINCIAL PERIODICALS

In addition to bird periodicals with a North American scope, state and provincial organizations offer regional publications for nearly all of North America. These are further supplemented by several hundred local bird journals and newsletters that describe birds and bird-watching activities near almost every major North American city. To find the bird club nearest your home, consult *A Guide to North American Bird Clubs* by Jon E. Rickert (see page 260).

Many regional and local periodicals are written by and for amateur bird-watchers. Articles and short field notes from serious amateurs comprise the majority of text for most publications in this category. Such periodicals provide an important

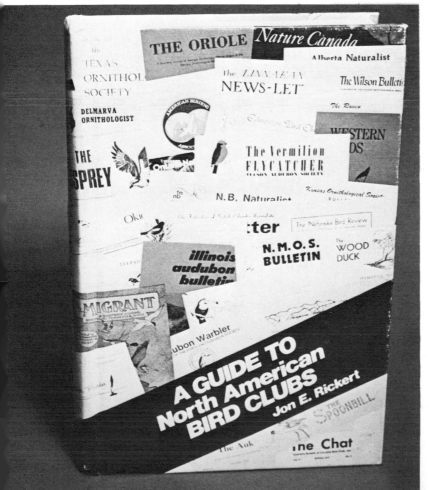

169. *A Guide to North American Bird Clubs* by Jon E. Rickert is the book to consult to find the bird club nearest to you.

service to the advancement of ornithology by giving amateurs a ready outlet for the publication of their field observations.

State Periodicals

Alabama Birdlife. Journal of the Alabama Ornithological Society, c/o Mr. J. Russel Bailey, 18 Peachtree Street, Birmingham, Ala. 35213.

Okiotak. Newsletter of the Anchorage Audubon Society, P.O. Box 1161, Anchorage, Alaska 99510.

The Road Runner. Newsletter of the Maricopa Audubon Society, c/o Dr. Robert Witzeman, 4619 E. Arcadia Lane, Phoenix, Ariz. 85108.

Arkansas Audubon Society Newsletter. Newsletter of the Arkansas Audubon Society, Museum of Science and Natural History, MacArthur Park, Little Rock, Ark. 72202.

Western Birds. Journal of the Western Field Ornithologists, 376 Greenwood Beach Road, Tiburon, Calif. 94920.

The Colorado Field Ornithologists Quarterly. Journal of the Colorado Field Ornithologists, P.O. Box 109, Berthoud, Colo. 80513.

The Connecticut Audubon Bulletin. Newsletter of the Connecticut Audubon Society, Main Office, 2325 Burr Street, Fairfield, Conn. 06430.

Delmarva Ornithologist. Journal of the Delmarva Ornithological Society.

Delmarva Ornithological Society Flyer. Newsletter of the Delmarva Ornithological Society, P.O. Box 4247, Greenville, Del. 19807.

Florida Field Naturalist. Journal of the Florida Audubon Society and Florida Ornithological Society.

Ornithological Newsletter. Newsletter of the Florida Audubon Society and Florida Ornithological Society, 1701 N.W. 24th Street, Gainesville, Fla. 32605.

The Oriole. Magazine of the Georgia Ornithological Society.

The Georgia GOShawk. Newsletter of the Georgia Ornithological Society, P.O. Box 38214, Atlanta, Ga. 30334.

Elepaio. Newsletter of the Hawaii Audubon Society, P.O. Box 22832, Honolulu, Hawaii 96822.

Portneuf Valley Audubon Society Newsletter. Newsletter of the Portneuf Valley Audubon Society, P.O. Box 4755, Pocatello, Idaho 83201.

SRAS Newsletter. Newsletter of the Snake River Audubon Society, 2245 Calkins Avenue, Idaho Falls, Idaho 83401.

Grizzly. Newsletter of the Golden Eagle Audubon Society, P.O. Box 968, Nampa, Idaho 83651.

The Prairie Owl. Journal of the Palouse Audubon Society, P.O. Box 3156, University Station, Moscow, Idaho 83843.

Illinois Audubon Bulletin. Journal of the Illinois Audubon Society.

Audubon Newsletter. Newsletter of the Illinois Audubon Society, P.O. Box 441, Wayne, Ill. 60184.

The Cardinal. Newsletter of the Indiana Audubon Society, Inc.

The Indiana Audubon Quarterly. Journal of the Indiana Audubon Society, Inc., Mary Gray Bird Sanctuary, R.R. #6, Connersville, Ind. 47331.

Iowa Bird Life. Journal of the Iowa Ornithologists' Union, c/o Mr. Peter C. Petersen, Editor, 235 McClellan Boulevard, Davenport, Iowa 52803.

The KOS Newsletter. Newsletter of the Kansas Ornithological Society, Museum of Natural History, University of Kansas, Lawrence, Kans. 66045.

The Kentucky Warbler. Journal of the Kentucky Ornithological Society, Biology Department, University of Louisville, Louisville, Ky. 40208.

LOS News. Newsletter of the Louisiana Ornithological Society, c/o Dr. David T. Kee, Editor, Department of Biology, Northeast Louisiana University, Monroe, La. 71209.

Maine Bird Life. Journal produced by Michael K. Lucey, Box 280, R.F.D. #3, Bangor, Maine 04401.

Guillemot. Newsletter of the Sorrento Scientific Society, c/o Mr. William Townsend, Editor, Box 373, Sorrento, Maine 04677.

Maine Audubon Quarterly. Journal of the Maine Audubon Society.

Maine Audubon Newsletter. Newsletter of the Maine Audubon Society, 118 U.S. Route 1, Falmouth, Maine 04105.

Maryland Birdlife. Journal of the Maryland Ornithological Society, Clyburn Mansion, 4915 Greenspring Avenue, Baltimore, Md. 21209.

The Massachusetts Audubon Newsletter. Newsletter of the Massachusetts Audubon Society, S. Great Road, Lincoln, Mass. 01773.

The Jack-Pine Warbler. Newsletter of the Michigan Audubon Society.

Michigan Audubon. Journal of the Michigan Audubon Society, 7000 N. Westnedge, Kalamazoo, Mich. 49001.

The Loon. Journal of the Minnesota Ornithologists' Union.

M.O.U. Newsletter. Newsletter of the Minnesota Ornithologists' Union, James Ford Bell Museum of Natural History, 10 Church Street S.E., University of Minnesota, Minneapolis, Minn. 55455.

The MOS Newsletter. Newsletter of the Mississippi Ornithological Society, c/o Museum of Natural Science, 111 N. Jefferson, Jackson, Miss. 39202.

The Bluebird. Journal of the Audubon Society of Missouri, c/o Mrs. Katherine Wade, 2202 Missouri Boulevard, Jefferson City, Mo. 65101.

The Accipiter Express. Newsletter of the Flathead Audubon Society, c/o Mr. Dan Sullivan, Yellow Bay, Bigfork, Mont. 59911.

Yellowstone Valley Flyer. Newsletter of the Yellowstone Valley Audubon Society, P.O. Box 1075, Billings, Mont. 59103.

Nebraska Bird Review. Journal of the Nebraska Ornithologists' Union, Inc., University of Nebraska State Museum, Lincoln, Neb. 68588.

The RRAS Newsletter. Newsletter of the Red Rock Audubon Society, P.O. Box 42944, Las Vegas, Nev. 89104.

The Pelican. Newsletter of the Lahontan Aududon Society, P.O. Box 2304, Reno, Nev. 89505.

New Hampshire Audubon. Newsletter of the Audubon Society of New Hampshire.

New Hampshire Audubon Annual. Journal of the Audubon Society of New Hampshire, 3 Silk Farm Road, Concord, N.H. 03301.

Records of New Jersey Birds. Journal of the New Jersey Audubon Society, 790 Ewing Avenue, Franklin Lakes, N.J. 07417.

N.M.O.S. Bulletin. Newsletter of the New Mexico Ornithological Society.

N.M.O.S. Field Notes. Journal of the New Mexico Ornithological Society, c/o Mr. John P. Hubbard, 2016 Valle Rio, Santa Fe, N. Mex. 87501.

The Kingbird. Journal of the Federation of New York State Bird Clubs, Inc.

New York Birders. Newsletter of the Federation of New York State Bird Clubs, Inc., 20 Drumlins Terrace, Syracuse, N.Y. 13224.

The Prothonotary. Newsletter of the Buffalo Ornithological Society, c/o Buffalo Museum of Science, Humboldt Parkway, Buffalo, N.Y. 14211.

The Chat. Journal of the Carolina Bird Club, Inc.

CBC Newsletter. Newsletter of the Carolina Bird Club, Inc., P.O. Box 1220, Tryon, N.C. 28782.

The Prairie Naturalist. Journal of the North Dakota Natural Science Society, Box 1672, Jamestown, N. Dak. 58401.

The GAAS Newsletter. Newsletter of the Greater Akron Audubon Society, c/o Mrs. Ann Biscan, 797 N. Firestone Boulevard, Akron, Ohio 44306.

The Bulletin of the CAS. Journal of the Cleveland Audubon Society, 2063 E. Fourth Street, Cleveland, Ohio 44115.

The CAS Newsletter. Newsletter of the Columbus Audubon Society, c/o Mr. Charles Wheeler, Editor, 134 W. Beechwold Boulevard, Columbus, Ohio 43214.

The TNA Bulletin. Newsletter of the Toledo Naturalists' Association, c/o Mrs. Beatrice J. Waterbury, 3 Ginger Hill Land, Toledo, Ohio 43623.

Bulletin of the Oklahoma Ornithological Society. Journal of the Oklahoma Ornithological Society.

The Scissortail. Newsletter of the Oklahoma Ornithological Society, c/o Jack D. Tyler, Editor, Bulletin, Biology Department, Cameron University, Lawton, Okla. 73505.

The Tulsa Scissortail. Newsletter, c/o Lois Rodgers, Editor, 428 S. 107th E. Avenue, Tulsa, Okla. 74128.

Cleveland County Audubon Society Newsletter. Newsletter of the Cleveland County Audubon Society, c/o Doris Watt, Department of Zoology, University of Oklahoma, Norman, Okla. 73019.

The Kite. Newsletter of Stephens County Audubon Society, c/o John Craythorne, P.O. Box 745, Duncan, Okla. 73533.

Oregon Birds. Journal of the Southern Willamette Ornithological Club, Box 3082, Eugene, Oreg. 97403.

The Membership News Bulletin. Newsletter of the Presque Isle Audubon Society, P.O. Box 1783, Erie, Pa. 16507.

Bulletin of ASWP. Newsletter of the Audubon Society of Western Pennsylvania, Beechwood Farms Nature Reserve, 614 Dorseyville Road, Pittsburgh, Pa. 15238.

Field Notes of Rhode Island Birds. Newsletter of the Rhode Island Ornithological Club, c/o Mr. Severyn Dana, 106 E. Manning, Providence, R.I. 02906.

The CAS Newsletter. Newsletter of the Columbia Audubon Society, P.O. Box 5923, Columbia, S.C. 29250.

The PAS Newsletter. Newsletter of the Piedmont Audubon Society, P.O. Box 2486, Spartanburg, S.C. 29302.

South Dakota Bird Notes. Journal of the South Dakota Ornithologists' Union, c/o Mrs. June Harter, Editor, P.O. Box 236, Highmore, S. Dak. 57345.

The Migrant. Journal of the Tennessee Ornithological Society, Cumberland Museum and Science Center, 800 Ridley Avenue, Nashville, Tenn. 37203.

The TOS Newsletter. Newsletter of the Texas Ornithological Society, Inc.

Bulletin. Journal of the Texas Ornithological Society, Inc., P.O. Box 19581, Houston, Tex. 77024.

The Stilt. Newsletter of the Bridgerland Audubon Society, 1722 Saddle Hill Drive, Logan, Utah 84321.

Vermont Institute of Natural Science Newsletter. Newsletter of the Vermont Institute of Natural Science.

Record of Vermont Birds. Newsletter of the Vermont Institute of Natural Science.

Vermont Natural History. Magazine (annual) of the Vermont Institute of Natural Science, Woodstock, Vt. 05091

The Raven. Journal of the Virginia Society of Ornithology.

The VSO Newsletter. Newsletter of the Virginia Society of Ornithology, c/o Mrs. John Dalmas, Treasurer, 520 Rainbow Forest Drive, Lynchburg, Va. 24502.

The Murrelet. Journal of the Pacific Northwest Bird and Mammal Society, c/o Dr. Richard E. Johnson, Editor, Department of Zoology, Washington State University, Pullman, Wash. 99194.

The Redstart. Magazine of the Brooks Bird Club, Inc.

The Mailbag. Newsletter of the Brooks Bird Club, Inc., Headquarters, 707 Warwood Avenue, Wheeling, W. Va. 26003.

Passenger Pigeon. Journal of the Wisconsin Society for Ornithology.

Badger Birder. Newsletter of the Wisconsin Society for Ornithology, Alex Kailing, Membership Chairman, W330 W8275 W. Shore Drive, Hartland, Wis. 52309.

Plains and Peaks. Newsletter of the Murie Audubon Society, P.O. Box 2112, Casper, Wyo. 82602

Provincial Periodicals

Alberta Naturalist. Journal of the Federation of Alberta Naturalists, Box 1472, Edmonton, Alberta 75J 2N5.

The FBCN Newsletter. Newsletter of the Federation of British Columbia Naturalists, P.O. Box 33797, Station D, Vancouver, British Columbia V6J 4L6.

Manitoba Naturalists Society Bulletin. Journal of the Manitoba Naturalists Society, Room 214, 190 Rupert Avenue, Winnipeg, Manitoba R3B 0N2.

The New Brunswick Naturalist. Journal of the New Brunswick Federation of Naturalists, c/o New Brunswick Museum, 277 Douglas Avenue, Saint John, New Brunswick E2K 1E5.

The NNHS Newsletter. Newsletter of the Newfoundland Natural History Society, P.O. Box 1013, St. John's, Newfoundland A1C 5M3.

The Nova Scotia Bird Society Newsletter. Newsletter of the Nova Scotia Bird Society.

The Fall News Bulletin. Newsletter of the Nova Scotia Bird Society, Nova Scotia Museum, 1747 Summer Street, Halifax, Nova Scotia B3H 3A6.

Ontario Naturalist. Journal of the Federation of Ontario Naturalists, Moatfield Park, 355 Lesmill Road, Don Mills, Ontario M3B 2W8.

The NHSPEI Newsletter. Newsletter of the Natural History Society of Prince Edward Island, P.O. Box 2346, Charlottetown, Prince Edward Island zip code TK.

Tchebec. Annual Report of the Province of Quebec Society for the Protection of Birds, Inc.

Newsletter. Newsletter of the Province of Quebec Society for the Protection of Birds, Inc., P.O. Box 43, Station B, Montreal, Quebec H3B 3J5.

Bulletin Ornithologique. Journal of the Club des Ornithologues du Quebec, Inc.

Feuille de Contact. Newsletter of the Club des Ornithologues du Quebec, Inc., 8191 de Zoo, Orsainville, Quebec G1G 4G4.

Blue Jay. Journal of the Saskatchewan Natural History Society.

Saskatchewan Natural History Society. Newsletter of the Saskatchewan Natural History Society, Box 1784, Saskatoon, Saskatchewan S7K 3S1.

The YCS Newsletter–Magazine. Newsletter of the Yukon Conservation Society, P.O. Box 4163, Whitehorse, Yukon Territory.

Rare Bird Alerts

Many Audubon groups and bird clubs offer up-to-date information about unusual bird sightings in their region. Known as bird "hot lines," "dial-a-bird," or "rare bird alerts," this popular service usually consists of about two minutes of prerecorded message about rare bird sightings and program activities. Consult *A Guide to North American Bird Clubs* by Jon E. Rickert (see page 260) for the phone numbers of local contacts who can provide the latest bird sightings and news. The following list of prerecorded telephone numbers is based on "Rare-Bird Alerts," by Noel J. Cutright (*Birding,* June 1980, volume XII, number 3).

Alaska
ANCHORAGE (ANCHORAGE AUDUBON SOCIETY)
(907) 694-3503

British Columbia
VICTORIA (VICTORIA NATURAL HISTORY SOCIETY)
(604) 478-8534

California
LOS ANGELES (LOS ANGELES AUDUBON SOCIETY)
(213) 874-1318

SAN FRANCISCO (GOLDEN GATE AUDUBON SOCIETY)
(415) 843-2211

SANTA BARBARA (SANTA BARBARA AUDUBON SOCIETY)
(805) 964-8240

Connecticut
(CONNECTICUT AUDUBON COUNCIL/DENISON PEQUOTSEPOS NATURE CENTER)
(203) 572-0012

Illinois
CENTRAL (ILLINOIS STATE MUSEUM)
(217) 785-1083

CHICAGO (CHICAGO AUDUBON SOCIETY)
(312) 675-8466

Maine
PORTLAND (MAINE AUDUBON SOCIETY)
(207) 781-2332

Maryland/Washington, D.C.

(AUDUBON NATURALIST SOCIETY OF THE CENTRAL ATLANTIC STATES)
(301) 652-1088

Massachusetts
BOSTON (MASSACHUSETTS AUDUBON SOCIETY)
(617) 259-8805

Michigan
DETROIT (DETROIT AUDUBON SOCIETY)
(313) 792-7140

Minnesota
MINNEAPOLIS (AUDUBON CHAPTER OF MINNEAPOLIS/MINNESOTA ORNITHOLOGISTS' UNION)
(612) 544-8315

New Hampshire
(NEW HAMPSHIRE AUDUBON SOCIETY)
(603) 224-9900

New Jersey
(NEW JERSEY AUDUBON SOCIETY)
(201) 766-2661

New York

ALBANY (HUDSON-MOHAWK BIRD CLUB)
(518) 377-9600

BUFFALO (BUFFALO MUSEUM OF SCIENCE)
(716) 896-1271

NEW YORK (LINNAEAN SOCIETY OF NEW YORK/NATIONAL AUDUBON SOCIETY)
(212) 832-6523

Ohio

CLEVELAND (KIRTLAND BIRD CLUB)
(216) 696-8186

COLUMBUS (COLUMBUS AUDUBON SOCIETY)
(614) 221-9736
(614) 882-8325 (Walden Pond–Columbus Metropolitan Park Board)

TOLEDO (TOLEDO NATURALISTS' ASSOCIATION AND MAUMEE VALLEY AUDUBON SOCIETY)
(419) 531-4420

Oregon

PORTLAND (PORTLAND AUDUBON SOCIETY)
(503) 292-0661

Pennsylvania

PITTSBURGH (AUDUBON SOCIETY OF WESTERN PENNSYLVANIA)
(412) 963-6104

Vermont

(VERMONT INSTITUTE OF NATURAL SCIENCE)
(802) 457-2779

Washington

SEATTLE (SEATTLE AUDUBON SOCIETY)
(206) 455-9722

Wisconsin

MILWAUKEE (WISCONSIN SOCIETY FOR ORNITHOLOGY)
(414) 352-3857

STATE AND PROVINCIAL AGENCIES

Although most state and provincial wildlife departments are concerned primarily with managing game species, an increasing number are employing nongame biologists and offering programs and materials to the public about nongame birds.

Most state and provincial natural resource agencies provide free or inexpensive publications about regional bird life. These often contain plans for constructing bird feeders, houses, and baths as well as tips for providing cover and food plantings especially adapted to local climatic conditions. Some agencies also maintain nurseries that specialize in growing wildlife food and cover plantings. Contact your state or provincial agency to see what bird-related materials are available.

Agents of the Cooperative Extension Service provide an additional source for publications about birds. The Extension Service maintains more than three thousand state and local offices throughout the United States. In addition to publications, extension agents, through 4-H Clubs, can provide educational programs about birds. They can also advise about attracting birds and solving nuisance problems. Contact your local agent or the state extension specialist who maintains an office at the nearest state land grant university. Locate your county extension agent by checking the telephone directory listings under U.S. Department of Agriculture or by writing to the Cooperative Extension Service, Washington, D.C. 20250.

State Wildlife Agencies

Department of Conservation and Natural Resources, 64 N. Union Street, Montgomery, Ala. 36130

Department of Fish and Game, Subport Building, Juneau, Alaska 99801

Game and Fish Department, 2222 W. Greenway Road, Phoenix, Ariz. 85023

Game and Fish Commission, Game and Fish Building, Little Rock, Ark. 72201

Department of Fish and Game, 1416 Ninth Street, Sacramento, Calif. 95814

Division of Wildlife, 6060 Broadway, Denver, Colo. 80216

Department of Environmental Protection, State Office Building, 165 Capitol Avenue, Hartford, Conn. 06115

Department of Natural Resources and Environmental Control, The Edward Tatnall Building, Legislative Avenue and William Penn Street, Dover, Del. 19901

Game and Fresh Water Fish Commission, 620 S. Meridian, Tallahassee, Fla. 32304

Department of Natural Resources, 270 Washington Street S.W., Atlanta, Ga. 30334

Division of Fish and Game, Department of Land and Natural Resources, 1151 Punchbowl Street, Honolulu, Hawaii 96813

Fish and Game Department, 600 S. Walnut Street, P.O. Box 25, Boise, Idaho 83707

Department of Conservation, 605 State Office Building, Springfield, Ill. 62706

Department of Natural Resources, 608 State Office Building, Indianapolis, Ind. 46204

State Conservation Commission, Wallace State Office Building, Des Moines, Iowa 50319

Forestry, Fish, and Game Commission, Box 54A, R.R. #2, Pratt, Kans. 67124

Department of Fish and Wildlife Resources, Capitol Plaza Tower, Frankfort, Ky. 40601

Department of Wildlife and Fisheries, 400 Royal Street, New Orleans, La. 70130

Department of Inland Fisheries and Wildlife, State Office Building, 284 State Street, Augusta, Maine 04330

Department of Natural Resources, Tawes State Office Building, Annapolis, Md. 21401

Division of Fisheries and Wildlife, 100 Cambridge Street, Boston, Mass. 02202

Department of Natural Resources, Stevens T. Mason Building, Box 30028, Lansing, Mich. 48909

Department of Natural Resources, Centennial Office Building, 658 Cedar Street, St. Paul, Minn. 55155

Game and Fish Commission, Robert E. Lee Office Building, 239 N. Lamar Street, P.O. Box 451, Jackson, Miss. 39205

Department of Conservation, P.O. Box 180, Jefferson City, Mo. 65101

Department of Fish and Game, Mitchell Building, 1420 E. Sixth, Helena, Mont. 59601

Game and Parks Commission, 2200 N. 33rd Street, P.O. Box 30370, Lincoln, Neb. 68503

Department of Fish and Game, Box 10678, Reno, Nev. 89520

Fish and Game Department, 34 Bridge Street, Concord, N.H. 03301

Division of Fish, Game and Shellfisheries, Department of Environmental Protection, Box 1809, Trenton, N.J. 08625

Department of Game and Fish, State Capitol, Santa Fe, N. Mex. 87503

Department of Environmental Conservation, 50 Wolf Road, Albany, N.Y. 12233

Wildlife Resources Commission, Archdale Building, 512 N. Salisbury Street, Raleigh, N.C. 27611

State Game and Fish Department, 2121 Lovett Avenue, Bismarck, N. Dak. 58505

Department of Natural Resources, Fountain Square, Columbus, Ohio 43224

Department of Wildlife Conservation, P.O. Box 53465, Oklahoma City, Okla. 73105

Department of Fish and Wildlife, P.O. Box 3503, Portland, Oreg. 97208

Game Commission, P.O. Box 1567, Harrisburg, Pa. 17120

Department of Natural Resources, 83 Park Street, Providence, R.I. 02903

Wildlife Resources Department, Building D, Dutch Plaza, P.O. Box 167, Columbia, S.C. 29202

Department of Game, Fish and Parks, Sigurd Anderson Building, Pierre, S. Dak. 57501

Game and Fish Commission, P.O. Box 40747, Ellington Building, Agricultural Center, Nashville, Tenn. 37204

Parks and Wildlife Department, 4200 Smith School Road, Austin, Tex. 78744

Division of Wildlife Resources, 1596 W. North Temple, Salt Lake City, Utah 84116

Fish and Game Department, Montpelier, Vt. 05602

Commission of Game and Inland Fisheries, 4010 W. Broad Street, P.O. Box 11104, Richmond, Va. 23230

Department of Game, 600 N. Capitol Way, Olympia, Wash. 98504

Department of Natural Resources, 1800 Washington Street E., Charleston, W.Va. 25305

Department of Natural Resources, Box 7921, Madison, Wis. 53707

Game and Fish Department, Cheyenne, Wyo. 82002

Provincial Wildlife Agencies

Fish and Wildlife Division, Department of Recreation, Parks, and Wildlife, 10363 108th Street, Sun Building, Edmonton T5J 1L7, Alberta

Fish and Wildlife Branch, Ministry of Recreation and Conservation, Parliament Buildings, Victoria V8V 1X4, British Columbia

Department of Mines, Resources, and Environmental Management, 310 Legislative Building, Winnipeg R3C 0V8, Manitoba

Fish and Wildlife Branch, Department of Natural Resources, Centennial Building, Fredericton E3B 5H1, New Brunswick

Department of Tourism, Wildlife Division, Building 810, Pleasantville, St. John's A1A 1P9, Newfoundland

Department of Natural and Cultural Affairs, Government of the Northwest Territories, Yellowknife X1A 2L9, Northwest Territories

Wildlife Division, Department of Lands and Forests, P.O. Box 698, Halifax B3J 2T9, Nova Scotia

Division of Fish and Wildlife, Ministry of Natural Resources, Whitney Block, 99 Wellesley Street W., Toronto M7A 1W3, Ontario

Fish and Wildlife Division, Environmental Conservation Services, Box 2000, Charlottetown C1A 7N8, Prince Edward Island

Fish and Game Branch, Department of Tourism, Fish, and Game, Place de la Capitale, 150 E. St. Cyrille Boulevard, Quebec City G1R 4Y1, Quebec

Department of Tourism and Renewable Resources, Fisheries and Wildlife, 1825 Lorne Street, Regina S4P 3N1, Saskatchewan

Yukon Territorial Government, Game Department, Box 2703, Whitehorse Y1A 2C6, Yukon Territory

8

BUILDING A BIRD-WATCHER'S LIBRARY

The growing popularity of bird-watching has created a flood of literature so great that it threatens to overwhelm the beginning bird enthusiast. For North America, there are dozens of books on identification, life histories, and bird finding. Each year new regional books appear, as well as coffee-table extravaganzas featuring bird photography and art.

This popularity is further evidenced by the leading college ornithology texts, most of which have recently appeared in new editions in an effort to keep pace with growing interest and advances in bird research. While excellent books exist about most bird-watching topics, there are also many inferior titles that confuse the selection process.

This chapter offers recommendations for building a basic bird-watcher's library. The books and sound recordings listed under the heading "Recommended" are those I feel will be most useful to beginning bird-watchers. Discussions of useful features and disadvantages follow the selected title entries. A goal for your bird library should be to include at least one recommended title from each of the following categories. At least one of the recommended entries is a low- or moderate-cost book.

The titles listed under "Other Choices" include entries that, in my opinion, are not basic to such a library, but are useful additions if resources permit. Some of the books under this heading are intended for advanced students.

Many public libraries contain a good selection of bird books. Use library books not only for short-term loans, but as an opportunity to carefully examine more expensive volumes before making purchases.

If the recommended books are not available at local bookstores, order them from retailers that specialize in bird books or directly from publishers. Many non-profit conservation organizations and nature centers sell books to support their programs, and they should be favored when possible. Library book sales and used-book dealers are likely places to locate out-of-print books. See Appendix B for a complete list of bird book retailers.

BIRD IDENTIFICATION

The best test of a field guide is its ability to permit quick and accurate field identifications under extremes of lighting and weather conditions. With positive identifications in mind, you can turn to more detailed treatments of life histories and behavior after returning home from the field. Field guides must be rugged enough to withstand harsh field conditions and be no bulkier than necessary. For this reason, hardback covers are preferable to paperbacks, as they hold up longer under field conditions.

Many birders pack a small field library in their cars, consisting of backup references to compare with their favorite field guide. Undoubtedly, one could learn to recognize North American birds from any of the field guides. The differences between the guides are unimportant when compared to the energy necessary to learn bird identifications. Despite differences in content and organization, there is no substitute for frequent use of your favorite guide. Thumb through the pages repeatedly—the more you glance over the plates, the more familiar the birds will become.

Recommended

Birds of North America. Chandler S. Robbins, Bertel Bruun, and H. S. Zim, with illustrations by Arthur Singer. New York: Golden Press, 1966. 340 pages, hardcover and paperback.

Birds of North America combines many convenient features to present all of North America's birds in one field-size volume. The book is strikingly illustrated by Arthur Singer's lifelike postured birds, which appear in traditional bird family organization with the most primitive birds in the front and the most advanced species toward the back. Picture keys and family descriptions permit the observer to first select the correct family and then to turn to the species descriptions. Typical silhouettes introduce each major bird group and provide comparisons with similar birds.

One-paragraph statements opposite the illustrations make it easy to move to the accompanying text. The text is clear and concise, giving specific identification tips about field marks and diagnostic behavior. The book includes sonograms for many of the more vocal species. These visual song "pictures" show frequency and pattern over time, but compared to good verbal descriptions they are awkward for most beginning birders. Abundant illustrations present the plumages of most females and immatures where they differ from adult males. Singer's illustrations provide much additional information regarding habitat and behavior simply by adding a few leaves or berries to the perches supporting the birds. He frequently includes nests or tucks a typical food item into a bird's beak where space permits as an aid to identification.

Although *Birds of North America* includes both eastern and western birds, a beginner can quickly eliminate birds out-of-range by glancing at the convenient range maps located opposite the illustrations. The color-keyed maps describe breeding ranges, winter distributions, and migration routes. Approximate spring arrival dates are also included for many species.

The presentation of eastern and western birds in the same volume has the advantage of exposing the reader to birds that would not be seen in a book with more regional limitations. Such exposure familiarizes the reader with additional species through association.

A Field Guide to the Birds, 4th edition. Roger Tory Peterson. Boston: Houghton Mifflin, 1980. (All birds east of the Rocky Mountains.) 384 pages, hardcover and paperback.

A Field Guide to Western Birds. Roger Tory Peterson. Boston: Houghton Mifflin, 1961. (All birds west of the Rocky Mountains, including the Hawaiian and Leeward islands.) 366 pages, hardcover and paperback.

A Field Guide to the Birds of Texas. Roger Tory Peterson. Boston: Houghton Mifflin, 1960. 304 pages, hardcover.

A Field Guide to the Birds, which first appeared in 1934, is widely credited with sparking the birth of popular bird-watching in North America. With several million copies sold, Peterson's field guides remain the standard introduction to bird identification. The Peterson bird identification system focuses on the use of field marks, identified in the illustrations by lines that point at distinctive features, such as patches of color or the shape of the neck, beak, or legs.

The wide acceptance of Peterson's eastern bird guide led to the production of western, Texas, Mexican, and European bird guides, all based on Peterson's popular field-mark system. The Peterson system is further expanded by twenty-four different natural history field guides on a variety of topics ranging from rocks and minerals to edible plants.

Peterson's bird guides have a conversational and enthusiastic quality that gives the text a personal tone. It is obvious that bird-watching is fun for Peterson, and he successfully shares his enjoyment with the reader. The guides provide verbal descriptions of song and range as well as comparisons with similar species.

The completely revised fourth edition contains color-coded range maps, but these are located in the back of the guide (to facilitate revisions) and are not as convenient as those in *Birds of North America.* Of real value to both the beginner and experienced birder are the many different plumages, which illustrate variety within each species. These are much more extensive than in any other field guide.

The Audubon Society Field Guide to North American Birds (Eastern Region). John Bull and John Farrand, Jr. New York: Alfred A. Knopf, 1977. 775 pages, softcover.

The Audubon Society Field Guide to North American Birds (Western Region). Miklos D. F. Volvardy. New York: Alfred A. Knopf, 1977. 825 pages, softcover.

These two guides contain an excellent, pocket-sized collection of North American bird photographs. The text is well written, containing useful species descriptions and accounts of habitat preferences with descriptions of nests and eggs. Verbal descriptions of bird songs are similar to those that appear in the Peterson guides and are easier to interpret than the sonograms featured in *Birds of North America.* The books include range maps in addition to range descriptions, but these are located in the text—not next to the photographs where they would be most convenient.

The books' principal shortcoming lies in their organization. Unlike the Robbins and Peterson guides, which organize birds according to family relationships, the Audubon Society guides place birds in arbitrary groups, such as "gull-like birds," "upland ground birds," and "swallow-like birds." The "perching birds," which comprise the largest category, are organized into groups of similar color, such as red birds, green birds, and brown birds. Although at first inspection this plan seems convenient, the system frequently places males and females in different sections of the book. Multicolored birds and birds of in-between color, such as greenish-brown or yellow-green, often have unpredictable positions. The book does little to prepare beginning bird-watchers to learn new bird families outside North America, where most other field guides are organized around knowledge of bird families.

The *Audubon Society Field Guides to North American Birds* are recommended principally for their excellent collection of bird photographs, which serve as a useful check against field identification assisted by either the Robbins or Peterson guides. The additional information on voice and nests along with the interesting general comment at the end of each entry make the guides useful additions to your library.

Other Choices

✓ *Cruickshank's Pocket Guide to the Birds: Eastern and Central North America.* Allan D. Cruickshank. New York: Simon and Schuster, 1972. Paperback.

Cruickshank gives suggestions for quick identification of bird families and provides excellent one-line suggestions for recognizing species. For example, he distinguishes between the hermit thrush and the wood thrush by noting that the hermit thrush is the only thrush with a reddish tail, while the wood thrush is the only thrush with reddish-brown on head, neck, and upper back.

A Field Guide to the Nests, Eggs and Nestlings of North American Birds. Colin Harrison. New York: William Collins Sons, 1978. 416 pages, hardcover.

Here is a valuable sequel to your field guide to adult North American birds. Species accounts describe the nests, eggs, and nestlings of all North American birds. Including interesting photos of behavior associated with nesting, it also contains photographs of North American bird eggs and illustrations of many nestlings.

A Field Guide to Birds' Nests Found East of the Mississippi. Hal H. Harrison. Boston: Houghton Mifflin, 1975. 257 pages, hardcover.

This field guide features excellent photographs of nests and eggs of 285 eastern species. The accompanying text describes breeding ranges and nesting habitats.

A Field Guide to Western Birds' Nests. Hal H. Harrison. Boston: Houghton Mifflin, 1979. 279 pages, hardcover.

This companion to the eastern bird nest guide presents discussion and photographs of nests and eggs for 520 species that breed west of the Mississippi.

The Habitat Guide to Birding. Thomas P. McElroy, Jr. New York: Alfred A. Knopf, 1974. 257 pages, hardcover.

The kinds of birds that live in various habitats are as predictable as the vegetation. This book provides descriptions of various habitats, such as hardwood and evergreen forests, brushy borders, roadsides, ponds, and seashores, and lists the birds you are likely to find in each habitat with details of the behavior and adaptations of representative species.

A Guide to Bird Songs, 2nd edition. Aretas A. Saunders. New York: Doubleday, 1951. 307 pages, hardcover.

Using paraphrases and line illustrations, Saunders interprets the songs of eastern land birds and shorebirds, providing novel ways to identify songs and calls. In field-guide format, *A Guide to Bird Songs* is a useful aid to learning bird songs and calls.

A Field Guide to the Seabirds of Britain and the World. G. Tuck and H. Heinzel. New York: William Collins Sons, 1978. 292 pages, hardcover.

Abundant color plates and line drawings provide illustrations of the approximately three hundred species of seabirds in the world. Published in field-guide format, with a water-resistant binding, this is a useful guide for birders who plan on voyaging the seas away from North America. Range maps indicate breeding range and known regions of occurrence for all species.

How to Know the Birds. Roger Tory Peterson. New York: New American Library, 1971. 144 pages, paperback.

This is an excellent review of the considerations important for making bird identifications. The book contains descriptions of many bird families and includes comments on the birds that frequent different habitats. Several pages of bird silhouettes test your ability to recognize bird shapes.

Birds. Herbert S. Zim and Ira N. Gabrielson. New York: Golden Press, 1956. 160 pages, hardcover and paperback.

This pocket-sized book contains descriptions of 129 of the most familiar North American birds, with comments on an additional 121 species. Attractive illustrations and easy-to-interpret range maps make this a useful introduction to common birds, but it is not extensive enough in scope to replace a standard field guide. *Birds* also contains a useful table listing arrival and departure dates and descriptions of eggs, nests, and food for 110 common species.

LIFE HISTORIES

After identifying birds in the field, it is important to do some reading once you return home. Make a point to read about the life histories of familiar birds and of new birds as you see them. Knowledge of family life, migrations, and other features of behavior add important depth to field experiences.

Recommended

Life Histories of North American Birds. Arthur C. Bent. Washington, D.C.: U.S. National Museum, 1919–68. Available as unabridged republication of the complete series by Dover Publications, New York. 23 vols., paperback.

This classic reference series contains a remarkable abundance of information about all North American birds. The books were compiled by Arthur Cleveland Bent under the auspices of the Smithsonian Institution—a project that took nearly fifty years to complete. In addition to Bent's own extensive observations, the series includes abundant narratives from many of Bent's contemporaries, as well as those from earlier naturalists, such as Audubon, Burroughs, and Brewster.

Each species' discussion contains such headings as nesting, eggs, young, plumages, food, and behavior, as well as spring and fall migration dates and observations. Bent succeeds in delivering all of this information in a highly readable, conversational manner.

A Guide to the Behavior of Common Birds. Donald W. Stokes. Boston: Little, Brown, 1979. 336 pages, hardcover

Long overdue, at last there is a readable summary of scientific literature describing the everyday habits of common birds. In field-guide format, this guide to "behavior watching" shows that bird behavior is as interesting as bird identification. By watching the body language of birds and listening for specific sounds, you can understand the basis of bird communication.

Instead of looking over a flock of mallards for "new" or rare species, use this guide to search the mallard's behavior repertoire. For each species, the book contains a full-page line drawing by the famed bird artist J. Fenwick Lansdowne and many posture sketches by the author. Each of the twenty-five species accounts contains a behavior calendar chart showing which behaviors to look for throughout the year, an illustrated selection of typical behaviors, and a discussion of the meaning of these behaviors. Written in conversational first-person format, the selected species accounts represent nineteen different families.

Audubon Land Bird Guide. Richard H. Pough. New York: Doubleday, 1949. 312 pages, hardcover.
Audubon Water Bird Guide: Water, Game and Large Land Birds. Richard H. Pough. New York: Doubleday, 1951. 352 pages, hardcover.
Audubon Western Bird Guide. Richard H. Pough. New York: Doubleday, 1957. 316 pages, hardcover (out of print).

This three-volume, field-guide format series can function in your library like a miniature Bent's Life History series. Although originally intended as a field identification guide, more recent guides better accomplish that function. The strength of the Pough series is the abundant text, which provides considerable information about the habits of North American birds. Each species entry not only includes descriptions of habits and methods of identification, but also a discussion of voice, nest, eggs, and range. Because of its field-guide size, the series can easily be packed in your car as part of a traveling field library. Excellent color illustrations by Don Eckelberry detail many plumages not found in other field guides.

Other Choices

Song and Garden Birds of North America. Alexander Wetmore et al. Washington, D.C.: National Geographic Society, 1964. 400 pages, hardcover.
Water, Prey, and Game Birds of North America. Alexander Wetmore et al. Washington, D.C.: National Geographic Society, 1965. 464 pages, hardcover (out of print).

These two volumes contain vibrant accounts of 656 species found in the United States, Canada, and parts of northern Mexico. Each chapter is written by an authority on a certain group of birds, and the result is a highly readable and authoritative text that is abundantly illustrated with art and photographs from dozens of North America's best nature artists and photographers.

A unique feature of both volumes is the inclusion of a small record album with excellent bird-song recordings. These albums present the songs and calls of 167 species with interesting commentary for each entry.

Studies in the Life History of the Song Sparrow. Margaret Morse Nice. New York: Dover, vol. 1, 1937; vol. 2, 1943. Vol. 1, 264 pages; vol. 2, 328 pages; paperback.

The result of eight years of careful observation, *Studies in the Life History of the Song Sparrow* is a model for future life-history monographs. Treating every detail of this common bird's behavior, development, song, and reproduction, Nice's work suggests how much remains unknown about even the most abundant species.

The Herring Gull's World. Niko Tinbergen. New York: Harper and Row, 1960. 255 pages, paperback.

The familiar herring gull is the focus of this classic life-history study. More than an in-depth view of this one species, *The Herring Gull's World* describes behavioral experiments that provide a broad insight into how birds perceive their world.

Vanishing Birds. Jack Halliday. New York: Holt, Rinehart and Winston, 1978. 296 pages, hardcover.

More than a documentation of extinction and decline, *Vanishing Birds* relates reasons for extinction to biological vulnerabilities and discusses what the losses mean to humans. Clearly written for public understanding, the book approaches extinction from a biological perspective, opening with a chapter on origin of species. Although extinction is a natural part of the speciation process, Halliday notes that human activities have greatly accelerated the process. Rather than document the decline of each of the approximately 130 extinct or vanishing birds, a few case histories are developed in depth, followed by a general discussion of vanishing birds in North America, New Zealand, Europe, and Australia. The book concludes with a discussion of some of the current conservation programs designed to assist endangered birds.

Extinct and Vanishing Birds of the World, 2nd revised edition. James C. Greenway, Jr. New York: Dover Publications, 1967. 520 pages, paperback.

This volume includes life histories for 132 kinds of extinct or vanishing birds. Authoritative accounts give the common name, current status, range, and physical description for each of the species and subspecies. Where possible, the accounts include abundant detail about history of the birds' present decline or causes for extinction.

The Birds of Canada. W. Earl Godfrey. Ottawa: National Museum of Canada, 1967. 428 pages, hardcover.

This book describes the 518 kinds of birds known to occur in Canada. It gives both English and French names for each species and includes sections detailing identification methods for birds viewed in the hand as well as in the field. For each species, the book describes preferred habitat, nesting biology (descriptions of nest, eggs, and incubation period), and North American distribution. The Canadian distributions are well illustrated by individual range maps. The book contains color illustrations of most species and occasional black-and-white detailed sketches of unique characteristics, such as beaks and feet.

Handbook of North American Birds, vol. 1: Loons through Flamingoes; vol. 2: Waterfowl, Part 1; vol. 3: Waterfowl, Part 2. Ralph S. Palmer, ed. New Haven: Yale University Press, vol. 1, 1962; vols. 2 and 3, 1976. Vol. 1, 567 pages; vol. 2, 521 pages; vol. 3, 560 pages; hardcover.

These thorough and authoritative volumes summarize existing knowledge about each species. The volumes contain detailed accounts of such topics as plumage, occurrence of hybrids, subspecies, and field identification methods. They also include discussions of vocalizations, distribution, preferred habitat, and arrival and departure dates, as well as extensive behavioral discussion. This series is for the serious bird-watcher who is looking for thorough species accounts and is willing to work through some technical accounts. This effort will be rewarded by abundant literature citations and the most detailed and up-to-date species summaries available.

The Audubon Illustrated Handbook of American Birds. Edgar M. Reilly, Jr. New York: McGraw-Hill, 1968. 524 pages, hardcover.

This book contains individual accounts of the 875 bird species known to occur north of Mexico. The birds of Hawaii and extinct birds are included as are those that have strayed into our region only a few times. The book is notable for containing species accounts of all birds in this area. Most species are illustrated by black-and-white photographs and line drawings, although occasionally this format is broken by full-page color photographs that are primarily decorative in function. A useful addition to the end of most species accounts is the "suggested reading" section which points the reader toward an important research-oriented paper or book about the species.

BIRD BIOLOGY

A bird biology book will provide a resource capable of answering questions about a wide range of ornithology topics, including anatomy, evolution, migration, and flight. The two recommended books provide highly readable answers to many of the most frequently asked bird questions.

Recommended

The Life of Birds. Joel Carl Welty. Philadelphia: W. B. Saunders, 1975. 623 pages, hardcover.

This is the most thorough and readable book available on bird biology. Now in its second edition, it presents a comprehensive overview of fourteen thousand research papers and books on topics ranging from the mechanics of flight to the origin of birds. Chapter topics include feathers, skeleton, courtship and mating habits, bird ecology, and migration. Welty illustrates the remarkable variety of bird adaptations with many examples gleaned from birds throughout the world. The lively, highly readable text gives abundant reason to marvel at the different ways birds adapt to their various habitats.

The Life of Birds is illustrated with black-and-white photographs and detailed line drawings. Abundant citations throughout the text direct the reader to the long list of references at the conclusion of the book.

Watching Birds. Roger F. Pasquier. Boston: Houghton Mifflin, 1977. 301 pages, hardcover and paperback.

This book treats a variety of ornithology topics, such as origin, feathers, anatomy, voice, migration, and winter habits, but clearly does not have the depth offered by Welty's *The Life of Birds. Watching Birds* is a high-school-level text and, as such, is an excellent introduction to the topics it approaches. According to the author, the book was written for the general public rather than for scientists or advanced students. For this reason the text does not include literature citations, and there is no bibliography. For people looking for an informative introduction to basic ornithology topics, this book is a good place to start.

Other Choices

The World of Birds. Gianfranco Bologna. New York: Abbeville Press, 1978. 256 pages, paperback.

Translated from the Italian, this readable book is packed with information from the research literature. It is abundantly illustrated by an impressive collection of paintings and photographs featuring birds from throughout the world.

Bird. Lois and Louis Darling. Boston: Houghton Mifflin, 1962. 261 pages, hardcover.

This refreshing introduction to bird biology has topics ranging from evolution to behavior and anatomy. Very clear descriptions and novel illustrations make potentially complex topics easy to understand.

Bird Study: An Introduction to Ornithology for Youth. R. A. Howard, Jr., J. W. Kelley, and J. Tate, Jr. Ithaca: Cooperative Extension, 1974. 42 pages, paperback.

Five student pamphlets written for intermediate-age students introduce the pursuit of bird-watching and detail activities about territoriality, mobbing, nest building, and nestling development. An accompanying leader's guide provides useful background so that any adult can advise students as they complete the activities.

The World of Birds. James Fisher and Roger Tory Peterson. New York: Crown Publishers, 1969. 183 pages, hardcover.

James Fisher, one of Britain's best-known ornithologists, contributed the authoritative text to this book, and Roger Tory Peterson, bird painter and author of many field guides, illustrated the pages with abundant and colorful art. The book is an excellent introductory ornithology text for the novice bird-watcher. Although not as thorough as Welty's outstanding *Life of Birds* or as comprehensive as Pasquier's *Watching Birds, The World of Birds* is an excellent overview of basic ornithology topics. The 1964 large-format edition contains two hundred maps detailing the world distribution of all bird families, a feature not found in any other book.

A Guide to Bird Watching. Joseph J. Hickey. New York: Dover Publications, 1975. 252 pages, paperback.

Although largely unchanged since the original 1943 edition, *A Guide to Bird Watching* remains an enthusiastic and useful introduction to field ornithology technique. Addressed to the beginning bird student, the book discusses such topics as bird migration, population fluctuations, and distribution. It also presents activities that amateurs can pursue—how to conduct a life-history study, how to publish field notes, and how to become a bird bander.

Birds—Their Life, Their Ways, Their World. Christopher Perrins. New York: Harry N. Abrams, 1976. 160 pages, hardcover.

This large-format book features superb illustrations by Ad Cameron. The text treats basic ornithology topics, such as evolution, anatomy, and behavior, for the beginning bird-watcher. A major chapter describes representative birds from a variety of habitats around the world. Chapters on migration and population biology are especially interesting, as they present experimental evidence reviewing several theories of bird orientation and population dynamics.

The Birds. Roger Tory Peterson and the Editors of *Life.* New York: Time Inc., 1963. Distributed by Little, Brown. 192 pages, hardcover.

An abundance of photographs and illustrations by Peterson bring the topics of flight, food habits, migration, and communication to life for those who have just begun to notice the birds around them. This is a good introduction to some of the basic topics in bird biology.

Ornithology in Laboratory and Field, 4th edition. Olin Sewall Pettingill, Jr. Minneapolis: Burgess Publishing, 1970. 524 pages, hardcover.

Since its first appearance in 1939 under the title *A Laboratory and Field Manual of Ornithology,* this book has been widely accepted as the standard manual for college laboratory courses in ornithology. The book now serves its original function as a laboratory manual and doubles as an excellent bird biology text. In addition to reference citations at the conclusion of each chapter, the book also contains extensive appendixes treating such topics as field methods, preparation of a scientific paper, and clutch size, incubation, and ages of fledging for North American birds. There are also extensive bibliographies providing references to such topics as life-history studies of North American birds and regional bird books of the world.

The Audubon Society Encyclopedia of North American Birds. John K. Terres. New York: Alfred A. Knopf, 1980. 1,109 pages, hardcover.

Here in one volume is a magnificent photographic collection of North American birds. More than 875 color photographs and 800 black-and-white artists' illustrations accompany approximately 6,000 different alphabetical listings about the biology and life histories of all 847 North American birds. Including topics such as brief biographies of notable ornithologists, bird diseases and parasites, as well as fossil birds, this is by far the most ambitious and extensive book of its kind. The highly readable text, drawn largely from scientific ornithological literature, is abundantly referenced to an extensive bibliography.

Fundamentals of Ornithology, 2nd edition. Josselyn Van Tyne and Andrew J. Berger. New York: John Wiley and Sons, 1976. 808 pages, hardcover.

This college text is intended for students with some knowledge of ornithology. The text covers the fundamental topics of bird biology, providing abundant facts and references to research on many topics, such as paleontology, plumage and molt, behavior, migration, and reproduction. Each chapter ends with an extensive list of references.

An Introduction to Ornithology, 3rd edition. George J. Wallace and Harold D. Mahan. New York: Macmillan, 1975. 546 pages, hardcover.

Although intended as an introductory text for college undergraduates, this is a useful beginning text for anyone interested in basic ornithology facts and concepts. Selected references follow each chapter, and there is an extensive literature section at the end of the book. Appendixes include a list of national ornithology organizations, description of techniques for preparing bird study skins, and lists of endangered and declining birds.

ATTRACTING BIRDS

The best approach to attracting birds is the manipulation of the habitat to meet their needs. To attract the greatest variety, provide food, water, shelter, and suitable nesting places. The recommended books in this section present the habitat approach to attracting birds. They not only contain plans and suggestions for building birdhouses, feeders, and baths, but, more important, they also contain specific plans for developing your own wildlife refuge. If successful, your backyard sanctuary will not only attract a greater variety of birds, but it will support a community of other creatures such as rabbits, toads, and butterflies.

Recommended

The Hungry Bird Book: How to Make Your Garden Their Haven on Earth. Robert Arbib and Tony Soper. New York: Taplinger Publishing, 1971. 126 pages, hardcover.

This is an excellent source book showing ways to make your yard and home more appealing to birds. *The Hungry Bird Book* is an entertaining yet authoritative manual that details ways to attract a greater variety of birds by directing your energies toward fulfilling bird needs. In addition to presenting plans for building traditional bird feeders and houses, the book gives practical suggestions for improving bird habitats. Additional nest sites, for example, can be created by pruning vegetation and modifying crevices in the walls of sheds and outbuildings. Much of the book's appeal comes from Robert Gillmor's delightful and often amusing illustrations. Although the book is intended primarily for eastern North America, the bird attraction principles involved make the book appropriate throughout North America. The book concludes with a useful list of birds that use nest boxes or visit feeding stations. Each entry describes nesting and/or feeding behavior and provides specific suggestions for attracting the species to your yard.

Songbirds in Your Garden, 3rd edition. John K. Terres. New York: Hawthorn Books, 1977. 301 pages, paperback.

This popular guide is an excellent choice for a basic bird-attracting reference. Its conversational first-person accounts combined with an abundance of information will answer most of the questions usually posed by backyard bird-watchers. In addition to chapters detailing how to build birdhouses, feeders, and baths, other chapters discuss ornamental plants attractive to birds (including prairie, desert, and western habitats), special techniques for attracting hummingbirds, and suggestions for hand feeding wild birds.

Key words, subtopics, and commonly asked questions are in the wide margins of the page, helping the reader locate information. Although there are relatively few illustrations in the book, the text is so readable that they are not missed.

Songbirds in Your Garden concludes with a set of tables packed with bird-attracting information, which includes a detailed planning design for planting a backyard wildlife sanctuary, descriptions of flowering and fruiting seasons for wildlife food plants, and food preferences and nesting data for common garden birds.

Gardening with Wildlife. National Wildlife Federation, Editors. Washington, D.C.: National Wildlife Federation, 1974. 190 pages, hardcover (available only from the National Wildlife Federation).

Chapters by many naturalists demonstrate how to manage your yard for more wildlife. Color photographs of backyard creatures ranging in size from aphid-eating ladybird beetles to raccoons illustrate the variety of animals that a well-planned yard will support. *Gardening with Wildlife* is an ecological approach to attracting birds. While it contains traditional attraction techniques such as bird feeders and houses, it is also a practical guide to shaping a habitat that will attract birds and a surprising variety of other animals. This is accomplished by providing lists of suggested plantings, designs for many types of backyard water supplies, and detailed plans for designing your green island.

Other Choices

Attracting Birds from the Prairies to the Atlantic. Verne E. Davison. New York: Thomas Y. Crowell, 1967. 252 pages, hardcover (out of print).

This useful bird-feeding reference is the product of the author's research into food habits and preferences of North American birds. Introductory chapters cover the usual basics of building feeders and houses, but the unique feature of this book starts with Chapter 4. Entitled "Birds—What They Eat, Where They Nest, and How to Attract Them," this chapter lists all birds of eastern North America and describes their nesting distribution and diet (frequently at some length). Davison also details specific ways to attract certain birds by providing plantings for food and nest sites. The book contains an alphabetical listing of wildlife food plants and lists the birds attracted by these plants.

Trees, Shrubs, and Vines for Attracting Birds. Richard M. Degraaf and Gretchin M. Witman. Amherst: University of Massachusetts Press, 1979. 194 pages, hardcover.

Plant your property with native vegetation that has the greatest value to birds. Certain ornamentals such as forsythia and hydrangea have little food and cover value in comparison to such native favorites as American cranberrybush and flowering dogwood. *Trees, Shrubs, and Vines for Attracting Birds* describes 162 plants (most native) that provide excellent food, cover, and/or nest sites in eastern North America. The book details how the proper combination of vegetation and landscaping will not only offer food and cover to birds, but will also give energy-saving insulation to your home. Each of the plant discussions includes full descriptions of size and fruiting potential as well as range, fruiting period, propagation, and a list of birds that use the plant.

Sources of Native Seeds and Plants. Soil Conservation Society of America, 7515 NE Ankeny Rd., Ankeny, Iowa 50021, 1979. 20 pages, paperback.

A list of wholesale and retail suppliers of native plants and their seeds.

A Complete Guide to Bird Feeding. John V. Dennis. New York: Alfred A. Knopf, 1976. 288 pages, hardcover.

This book provides basic techniques for setting up a feeding station and gives suggestions for solving feeder problems, such as nuisance squirrels and depredations by neighborhood cats and dogs. It has little information about using plantings to provide bird habitat, cover, or feed. The book consists primarily of species accounts describing the behavior of eastern North American feeder birds and the approaches you can use to attract the greatest variety.

The Backyard Bird Watcher. George H. Harrison. New York: Simon and Schuster, 1979. 284 pages, hardcover.

In this personal account, Harrison shows how he and many other backyard conservationists are creating ideal bird habitats—often within large cities. The book provides model landscaping plans to attract birds and discusses feeding stations, housing, pest problems, and water supplies.

Bird Feeding Manual. Stephen W. Kress. Syracuse: Agway, 1973. 47 pages, paperback. Available from Cornell Laboratory of Ornithology.

This concise booklet offers tips for setting up and maintaining a backyard feeding station, contains plans for constructing homemade feeders, and offers suggestions for solving feeder-related problems, such as bird collisions at windows and control of squirrels and nuisance birds. A useful feature of the booklet is the checklist and calendar graph section that shows the months of occurrence and preferred foods for the feeder birds of the northeastern United States.

American Wildlife and Plants: A Guide to Wildlife Food Habits. Alexander C. Martin, Herbert S. Zim, and Arnold L. Nelson. New York: McGraw-Hill, 1951. Available as unabridged republication of the 1951 edition from Dover Publications, New York (1961). 500 pages, paperback.

This book summarizes extensive research by the U.S. Fish and Wildlife Service into the food habits of North American wildlife. Entries for more than three hundred common North American birds are included. For each species, summer and winter ranges within the United States are shown on range maps, and informative graphs show the relative proportion of plant and animal food consumed at each season of the year. For each species, there are also lists of preferred plants. Other chapters list the birds that feed on woody plants, upland weeds, and aquatic and cultivated plants. For most plants there are range maps and lists of the animals that consume their seeds and fruits. This classic reference to wildlife food habits does not stop with birds—it also describes the food habits of many North American mammals as well as a few fish, reptiles, and amphibians.

The New Handbook of Attracting Birds, 2nd edition. Thomas P. McElroy, Jr. New York: Alfred A. Knopf, 1975. 262 pages, hardcover.

This introduction to the basics of attracting birds provides theory about bird numbers and the factors that affect their abundance. The book presents construction plans for feeders, birdbaths, and nesting boxes as well as evaluations of plantings for attracting birds.

The Bluebird. Lawrence Zeleny. Bloomington: Indiana University Press, 1976. 170 pages, paperback.

There are few birds as responsive to human assistance as the three species of North American bluebirds. This is a thorough discussion of bluebird behavior and the environmental factors that affect their populations. On the premise that population declines are linked largely to a rapidly dwindling supply of natural nesting cavities, Zeleny details designs for building bluebird nest boxes and demonstrates how to maintain them. If you are interested in bluebirds and their future, this book is a must for your library.

BIRD-WATCHING ACTIVITIES

SELECTING BINOCULARS
"Binoculars." *Consumer Reports.* March 1980: 196–203.
Koehler, H. "Answers to Your Questions on Binoculars." *Birding.* March–April 1978: 88–91.
Reichert, R. J., and Elsa Reichert. *Know Your Binoculars.* 1951. Available from Mirakel Optical Co., 331 Mansion St., West Coxsackie, N.Y. 12192.

COUNTING BIRDS
Arbib, R. S. "On the Art of Estimating Numbers." *American Birds.* 26(3)(1972): 706–12.
Hickey, J. J. "Adventures in Bird Counting." From *A Guide to Bird Watching.* New York: Dover, 1975. 252 pages.
"Reprints from *American Birds* About Counting and Estimating Bird Populations."
> "The Audubon Winter Bird-Population Study," by Haven Kolb; "Breeding-Bird Censuses—Why and How," by George A. Hall; "Breeding-Bird Census Instruction," by Willet T. Van Velzen, "An Appraisal of the Winter Bird-Population Study Technique," by Chandler S. Robbins; "On the Art of Estimating Numbers," by Robert Arbib. To order and determine the current cost and availability of the above publications write: *American Birds,* National Audubon Society, 950 Third Ave., New York, N.Y. 10022.

TAKING FIELD NOTES
Herman, S. G. *The Naturalist's Field Journal: A Manual of Instruction Based on a System Established by Joseph Grinnell.* Olympia, Wash.: The Evergreen College Bookstore. 200 pages.
Remsen, J. V. "On Taking Field Notes." *American Birds.* 31(5)(1977): 946–53.

RECORDING BIRD SOUNDS

Borror, D. J. *Bird Song and Bird Behavior.* New York: Dover, Twelve-inch, 33⅓
   rpm monaural record and 32-page manual (see page 298)
Bradley, R. "Making Animal Sound Recordings." *American Birds.* 31(3)(1977):
   279–85.
Davis, T. H. "Microphones and Headphones for Bird-recording." *Birding.*
   11(5)(1979): 240–43.
———. "Cassette Tape Recorders Update." *Birding.* 10(4)(1978): 185–92.
Fisher, J. B. *Wildlife Sound Recording.* London: Pelham Books, 1977. 173 pages.
Gulledge, J. L. "Recording Bird Sounds." *The Living Bird.* 15(1976): 183–203.
King, Ben. "The Magic Wand" (shotgun microphones). *Birding* 12(3)(1980):
   106–8.

PHOTOGRAPHING BIRDS

Allen, A. A. *Stalking Birds with Color Camera.* Washington, D.C.: National
   Geographic Society, 1961. 328 pages.
Blaker, A. A. *Field Photography.* San Francisco: W. H. Freeman, 1976. 451 pages.
Bunker, H. L. "The Search for the Perfect Bird Lens Goes On." *American Birds.*
   31(6)(1977): 1083–86.
Cruickshank, A. D. *Cruickshank's Photographs of Birds of America.* New York: Dover,
   1977. 182 pages.
Linton, D. *Photographing Nature.* Garden City, New York: The Natural History
   Press, 1964. 262 pages.
Porter, E. *Birds of North America.* New York: Dutton, 1972. 144 pages.

BIRD CLUBS AND LOCAL CONTACTS

Next time you plan a trip, contact local birders near your intended destination;
they may be able to alert you to the presence of rarities and point the way to the
most productive local habitats. You'll soon find that most bird enthusiasts are
among the friendliest people you could meet anywhere.

Recommended

A Guide to North American Bird Clubs. Jon E. Rickert. Elizabethtown, Ky.: Avian
Publications, 1978. 564 pages, hardcover (available only from Avian
Publications, Inc., P.O. Box 310, Elizabethtown, Ky. 42701).
   To find the bird club nearest your town or trip destination, check the outline
maps of each state or province in this book, since these indicate numbered locations for each club. The book provides an astounding amount of information for
more than eight hundred clubs, including a contact address, description of publications, details about field trips and meetings, and one or more phone numbers

to call to report or learn about unusual birds. Some of these numbers are tape-recorded "dial-a-bird" messages that can be reached twenty-four hours a day. Frequent updates appear in *Birding News Survey* (see page 227).

Nature Guide. Ilene Marcks, Editor. Tahoma, Wash.: Tahoma Audubon Society, 1978–79 edition. 76 pages, paperback.

The most enjoyable and efficient way to find birds in a new region is with the assistance of a local guide. *Nature Guide* provides a list of volunteer naturalists throughout North America (and a few foreign countries) who offer their local knowledge to traveling nature enthusiasts. Many of the guides are competent not only with birds and other wildlife, but they are also familiar with botany and the geological history of their home region. This book also lists North American nature centers, complete with addresses and phone numbers, national wildlife refuges, National Audubon Society sanctuaries, and national parks and forests in the United States and Canada. North American maps show the location of national parks and wildlife refuges. *Nature Guide* also contains a list of some favorite bird-finding sites and bird-finding references for each state and province.

BIRD FINDING

The lure of seeing new bird habitats can add a special excitement to planning vacations and other travels. Try a birding vacation—it's certain to provide memorable experiences in out-of-the-way places.

Recommended

A Guide to Bird Finding East of the Mississippi, 2nd edition. Olin Sewall Pettingill, Jr. New York: Oxford University Press, 1977. 689 pages, hardcover. Paperback by Houghton Mifflin.
A Guide to Bird Finding West of the Mississippi. Olin Sewall Pettingill, Jr. New York: Oxford University Press, 1953. 709 pages, hardcover (out of print, 2nd edition in preparation).

Sparked by his own frustrations in tracking down bird-finding locales, Pettingill decided to produce a book that would direct bird enthusiasts to some of the best bird habitats in North America. Since the appearance of the first edition, there have been impressive changes in the distributions of many North American birds. Even more staggering is the loss of much excellent bird habitat to housing developments, shopping centers, and highways. These changes called for the greatly revised and enlarged edition of the eastern guide. A similar revision of Pettingill's western guide is in preparation. The new edition features updated travel directions and lists bird-finding sites (such as national, state, and private parks and refuges) that are not likely candidates for development.

Each chapter begins with an ecological overview of the significant bird habitats within a state. Lists of characteristic species follow each habitat description. In addition to parks and refuges, the book includes other relatively "permanent" bird habitats such as college campuses, airports, and cemeteries.

Guide to the' National Wildlife Refuges. Laura and William Riley. New York: Anchor Press/Doubleday, 1979. 653 pages, hardcover.

This guide provides detailed accounts for 180 U.S. national wildlife refuges with notes on the remaining 380 refuges. This thorough book tells which refuges are open to the public, how to get there, what to see and do, where visitors can stay or camp, the best times to visit, and much more. Organized into ten regions, each regional account provides an overview of the most notable refuges followed by a list of distinctive birds for the region. Easy-to-read state maps show the location of refuges in reference to nearby cities.

Other Choices

A Birdwatcher's Guide to the Eastern United States. Alice A. Geffen. New York: Barron's Woodbury, 1978. 346 pages, paperback.

This is a travel directory to parks and refuges in the eastern United States. For each state there is a general description of notable bird habitats and mention of some of the characteristic birds. This is followed by detailed descriptions of national bird-watching preserves found in the state, such as national parks and refuges. Nature centers and sanctuaries operated by Audubon societies and nature conservancy are noted.

Eastern Hawk Watching. Donald S. Heintzleman. University Park, Pa.: Keystone Books, Pennsylvania State University Press, 1976. 99 pages, hardcover and paperback.

This book contains descriptions, access directions, and suggested time of year to visit fifty-six hawk-watching sites in the eastern United States. In addition to notes on how to identify twenty species of hawks, vultures, and eagles, the book also contains flight photographs (and some illustrations) for twenty-five species.

Birdwatcher's Guide to Wildlife Sanctuaries. Jessie Kitching. New York: Arco Publishing, 1976. 233 pages, paperback.

With an enthusiastic tone, this book describes the location of 295 wildlife sanctuaries throughout North America and Canada. Comments about the availability of local bird lists are especially useful.

The Bird Watcher's America. Olin Sewall Pettingill, Jr., Editor. New York: Thomas Y. Crowell, 1974. 441 pages, paperback.

Bird Watcher's America offers a delightful vicarious birding experience at forty-four notable bird-finding sites in the United States and Canada. It also contains travel details and local information that will prove useful to those who actually visit the sites. Leading authorities describe each location—readers follow Roger Tory Peterson to the Pribilof Islands of Alaska, Herbert Krause into the Black Hills of South Dakota, and Allan D. Cruickshank to the Bird Islands of Maine.

Directory of Nature Centers and Related Environmental Facilities. New York: National Audubon Society, 1979. 329 pages, paperback.

Nature Centers and Outdoor Education Centers protect more than eight hundred seventy-four thousand wildlife habitats in the United States and Canada. These facilities, often located within or near large cities, are usually excellent bird-watching habitats and frequently offer interpretive programs dealing with birds and other wildlife. This directory provides addresses and a brief description about the programs of approximately one thousand nature centers and similar facilities in North America.

NORTH AMERICAN AND WORLD CHECKLISTS

Recommended

A.B.A. Checklist: Birds of Continental United States and Canada. Checklist Committee of the American Birding Association. Austin: The American Birding Association, 1975. 64 pages, paperback.

This is the official North American checklist of the American Birding Association. The *A.B.A. Checklist* lists the 794 species known to occur in the United States and Canada. The list does not include Bermuda, Hawaii, or Greenland. This checklist is revised each year by a supplement that appears in the November–December issue of *Birding.* Extra pages in the back of the checklist provide places to record additional species and modifications to the list.

The order in which the birds are listed and the names of most species largely follow the authoritative 5th edition of the *A.O.U. Checklist of North American Birds* (with supplements). However, some English names are modified in the A.B.A. list in an effort to "standardize and clarify" certain common names.

Traveler's List and Check List for Birds of North America. James A. Tucker. Austin: The American Birding Association, 1975. 28 pages, paperback.

The traveling bird-watcher will find this list of North American birds a convenient way to maintain a daily list. Each bird name is followed by eleven blank squares. The squares can be used for eleven different days or reserved for different locales. This list includes most species that regularly occur north of the Mexican border. Organization follows the *A.B.A. Checklist.* Space is also provided to list accidentals, subspecies, color phases, and other unusual sightings.

Birds of the World: A Check List, 2nd edition. James F. Clements. New York: Two Continents Publishing Group, 1978. 524 pages, hardcover.

This is a conveniently organized list of the world's birds. A total of 9,058 species are grouped into orders and families and identified by Latin and English names. Next to each entry, the observer will find space to record the date and location for their first field observation. A useful feature is the brief statement about the world range of each species. Birds that are unique to a particular locale (endemics) or those that are rare or endangered are identified as such in the text.

Other Choices

The North American Birder's Library Lifelist. Susan Roney Drennan, Editor. Garden City, N.Y.: Doubleday, 1979. 630 pages, hardcover.

Weighing in at $5\frac{1}{2}$ pounds, this is not a book for use anywhere but in your library! For each of 821 species, there is space to record details of sightings—date first seen, location, habitat, time, weather, and other observations. A grid of state and province blocks gives a place to record these scores as well as yearly totals, big day counts, Christmas counts, and more.

The Birder's Field Notebook. Susan Roney Drennan, Editor. Garden City, N.Y.: Doubleday, 1979. 158 pages, paperback.

This pocket-sized book provides a standard form for recording field descriptions of unknown birds. The categories provide space to record details of color and form that will be useful in making identifications on return from the field.

Checklist of the World's Birds. Edward S. Gruson. New York: Quadrangle/New York Times Book Company, 1976. 212 pages, hardcover.

This list of world birds provides one or more references for each species by conveniently placing code symbols next to the species name. A similar code system also identifies the bird's worldwide distribution. Although the book is "intended for listers and tickers," there is no space reserved for the birder to record field notes. For this reason the book will not be as useful as the Clements list of world birds to the traveling birder who wants to keep a cumulative record of birds encountered.

REGIONAL AND LOCAL REFERENCES

Your bird library should contain at least one reference about the birds that inhabit your home region. Such books and checklists usually indicate which birds to expect in appropriate habitats and offer information about abundance, distribution, and seasons of occurrence. Before traveling to new bird habitats, consult a local checklist or bird-finding guide to familiarize yourself with the birds you are likely to see. Preparation makes a big difference.

The following selection contains most of the state and provincial titles as well as local checklists for some of the most popular bird-watching sanctuaries. The list includes book reviews as noted by Susan Roney Drennan (SRD)* and James A. Tucker (JAT).** Because some of the publications have limited distributions,

*Susan Roney Drennan. "Annotated Selection of Regional Bird Books." *American Birds* 30 (October 1976): 2–20.

**James A. Tucker, ed. Supplement to *Birding* 8 (September–October 1976).

it is best to send a letter of inquiry to the publisher or indicated source to learn about availability and current price before placing your order. When ordering field checklists, send a stamped, self-addressed, legal-sized envelope with your inquiry.

United States

ALASKA

The Birds of Alaska. Ira N. Gabrielson and Frederick C. Lincoln. Stackpole Co. and The Wildlife Management Institute, 1959. 922 pages. Available from Stackpole Books, Cameron and Keller Streets, Harrisburg, Pa. 17105. A thorough state reference containing history of Alaskan ornithology, species range in Alaska, abundant records and field notes, list of specimens in Russian collections, "habits and haunts" discussion, gazetteer, bibliography, ecological zones explanation, 13 four-color plates, and an appendix of species added through July 1958. Parts are dated; a valuable source. (SRD)

Status and Distribution of Alaska Birds. Brina Kessel and Daniel D. Gibson. 1979. 100 pages. Available from ABA Sales. This is an update of Gabrielson and Frederick's *Birds of Alaska.* Complete accounts are included only for species not adequately treated in the original title. (JAT)

Checklist of Birds of Alaska. D. D. Gibson. 1977. 5 pages. College of Biological Sciences and Renewable Resources, 101 Irving Building, University of Alaska, Fairbanks 99701. Also available from ABA Sales, P.O. Box 4335, Austin, Tex. 78765.

Birding the 49th State: An Alaskan Saga. Harold Morrin. 1978. 130 pa.ges. Available from ABA Sales. Informative and entertaining. (JAT)

Alaska Bird Finding. Hall. Reprinted from *Okiotak* by Anchorage Audubon Society, P.O. Box 1161, Anchorage 99510. Eight articles on bird finding in Alaska.

Birds of Mt. McKinley National Park. Olaus Johon Murie. 1963. 86 pages. Alaska National Parks and Monuments Association, P.O. Box 2252, Anchorage 99501. Introduction describes park. Species descriptions give good idea of status and relative abundance in park, as well as best areas. Index and checklist. (JAT)

Birds of Southeast Alaska: A Checklist. 1978. 24 pages. Department of Agriculture, Forest Service, Alaska Region, P.O. Box 1628, Juneau, Alaska 99802. Status, relative abundance by habitat for southeastern Alaska and the coast range. Introduction, lists of notable birds of Upper Glacier Bay, birds of open sea, and marine mammals of inside waters (with notes). Nesting, bibliography. (JAT)

ARIZONA

Birds of Arizona. Allan Phillips, Joe Marshall, and Gale Monson. 1964. 212 pages. University of Arizona Press, P.O. Box 3398, Tucson 85722. An excellent, illuminating, scholarly treatment. The range maps are especially well executed. Considerable attention to taxonomy. (SRD)

Field Checklist of Arizona's Birds. Richard L. Todd. 1976. 8 pages. Arizona Game and Fish Department, 2222 W. Greenway Road, Phoenix 85023.

Birds in Southeastern Arizona, 2nd ed. William A. Davis and Stephen M. Russell. 1980. 126 pages. Tucson Audubon Society, P.O. Box 40115, Tucson 85717. Directions and comments for 43 top birding spots, many not cited elsewhere. Bar graphs indicating abundance, with habitat notes, are included for 450 species. Covers Arizona and western Mexico.

Birder's Guide to Southeastern Arizona. James A. Lane. 1974. 110 pages. L and P Press, Box 21604, Denver, Colo. 80221. Also available from ABA Sales. A travel guide designed especially for the vacationing bird-watcher. Precise mileage and directions are given for finding the best birding spots. General information on the area and lists of birds and mammals giving status, relative abundance, and habitats. Maps and directions for several trips. (JAT)

Bird-finding Localities in Southwestern New Mexico and Southeastern Arizona. Manager, San Andres National Wildlife Refuge, Box 756, Las Cruces, N. Mex. 88001.

Birds of Maricopa County, Arizona. Salome R. Demaree et al. 1972. 70 pages. Maricopa Audubon Society Available from Mrs. Robert Witzeman, 4619 E. Arcadia Lane, Phoenix 85018. Introduction. Descriptions and directions for 13 birding areas, plus 4 areas for special birds. Bar graphs, nesting. Notes for each species give such information as status, abundance, habitat, dates of rarities, and documentation information. Map, index, notes section. (JAT)

Birds of Organ Pipe Cactus National Monument. Richard A. Wilt. 1976. 82 pages. Southwestern Parks and Monuments Association, Box 1562, Globe, N. Mex. 85501. General description of region, including habitats and birding areas. Annotated list gives distribution, abundance, habitat, breeding. Photos. (JAT)

ARKANSAS

Birds of Arkansas, 2nd revised edition. W. J. Baerg. 1951. 258 pages. Agricultural Experiment Station, University of Arkansas College of Agriculture, Fayetteville 72701. Adequate but not exceptional treatment of Arkansas birds. Needs updating but will serve usefully until a new edition appears. (SRD)

Arkansas Audubon Society Field List. 1974. 4 pages. Arkansas Audubon Society, 5809 N. Country Club, Little Rock 72207.

Birds of Northeastern Arkansas. Earl L. Hanebrink. 1980. 48 pages. Stuart Rockwell, publisher. Available from the author, Box 67, State University 72467. A habitat-oriented summary of 15 years of bird observations. Includes seasonal bar graphs for all species, results from breeding bird surveys, censuses, and Christmas counts. Also descriptions of 17 good birding sites, and bibliography.

CALIFORNIA

Birds of California. Arnold Small. 1974. 310 pages. Winchester Press, 205 E. 42nd Street, New York, N.Y. 10017. Issued in paperback by Collier Macmillan, 866 Third Avenue, New York, N.Y. 10022. Small describes 25 different habitats and their bird denizens. An annotated list of the 518 birds recorded in California gives habitat, seasonal status, and range. More than 300 photographs. The reader will long for a map indicating localities mentioned in the text. There is no botanical reference, no bibliography, no geological reference. (SRD)

Handbook of California Birds, 2nd edition. Vinson Brown, Henry Weston, Jr., and Jerry Buzzell. 1973. 224 pages. Naturegraph Publishers, Inc., P.O. Box 1075, Happy Camp 96039. A revised and enlarged edition, but not truly a monumental, weighty handbook. Of the more than 530 species recorded in California, 368 are briefly described here and illustrated in color. There is a generous sprinkling of black-and-white line drawings throughout to assist with identification and study in the field. There are separate chapter sections on flight, courtship, foraging, plumages, bill and foot function, eggs and nests, and conservation. Unfortunately, the bird names contain several inconsistencies owing to lack of name updating. However, the illustrations, especially the line drawings of Buzzell, deserve special mention, and in general the handbook will be useful to the student of California avifauna. (SRD)

Los Angeles Audubon Society Field List of the Birds of California. Guy McCaskie. 1978. 18 pages. Los Angeles Audubon Society, 7377 Santa Monica Boulevard, Plummer Park, Los Angeles 90046. Checklist of California birds.

Birder's California. Don Robertson. 1978. 108 pages. Available from ABA Sales. Thirty-five maps to California birding locations.

Checklist of Northwest California Birds: Del Haste, Humboldt, Northern Mendocino, Western Trinity and Western Siskiyou Counties. S. W. Harris. 1978. 6 pages. Redwood Region Audubon Society, Box 1054, Eureka 95501.

Birds of Northern California. Guy McCaskie et al. 1979. 84 pages. Golden Gate Audubon Society, 2718 Telegraph Avenue, Suite 206, Berkeley 94705. Bar graphs showing arrival and departure dates and extensive notes on field identification.

Birds of Yosemite. Cyril and Robert C. Stebbins. 1967. 152 pages. Yosemite Natural History Association, Yosemite National Park 95389. Introduction includes description of habitats and life zones, with plant and bird indicators for life zones of Yosemite. Detailed descriptions and many drawings of species. Silhouettes of vultures, hawks, and eagles. Detailed species accounts include information on location in park. List of rarities. Appendixes: I. Family Characteristics of Birds (head and foot drawings of representative species); II. Key to Yosemite Birds. Index and checklist with status and life zones. (JAT)

Birds of the Yosemite Sierra. David Gaines. 1977. 154 pages. Available from ABA Sales. An annotated checklist of western and eastern slopes of the Sierra.

A Checklist of Birds to be Found Within Fifty Miles of the City of Santa Barbara, California. Richard Webster. n.d. 8 pages. Santa Barbara Museum of Natural History, Museum Shop, 2559 Puesta del Sol, Santa Barbara 93105. Field checklist providing seasonal abundance and habitat designations.

The Birds of Santa Barbara and Ventura Counties, California. Richard Webster et al. 1980. 43 pages. Santa Barbara Museum of Natural History Occasional Paper No. 10. Museum Shop, 2559 Puesta del Sol, Santa Barbara 93105. Map of both counties with bar graphs showing seasonal abundance for 423 species. Annotated notes providing dates of occurrence are included for many of the species.

Productive Birding Spots in San Diego County, California. Claude G. Edwards, Jr. 1980. 4 pages. Claude G. Edwards, Jr. San Diego Field Ornithologists, 2932 Greyling

Drive, San Diego 92123. Map and key to good birding sites with a list of rare birds. Site guides in preparation for 37 locations within the county.

Birder's Guide to Southern California, revised and enlarged edition. James A. Lane. 1979. 140 pages. L & P Press, Box 21604, Denver, Co. 80221.

The Bird Year: A Book for Birders with Special Reference to the Monterey Bay Area. John Davis and Alan Baldridge. 1980. 224 pages. The Boxwood Press, 183 Ocean View Blvd., Pacific Grove 93950. Bird habitats, local history, bibliography, and bird list.

Pelagic Birds of Monterey Bay, California. Richard Stallcup. 1976. 24 pages. Available from ABA Sales.

List of the Birds of the Monterey Peninsula Region. Laidlaw Williams with additions by the checklist committee. 1977. 8 pages. Pacific Grove Museum of Natural History, Pacific Grove 93950.

Birds of the Lake Tahoe Region. Robert T. Orr and James Moffitt. 1971. 150 pages. California Academy of Sciences, Golden Gate Park, San Francisco 94118. Historical summary, major plant associations, and extensive species accounts. Photographs, references, and index. (JAT)

COLORADO

Birds of Colorado. Alfred M. Bailey and Robert J. Niedrach. 2 vols., illustrated. 1965. 895 pages. Denver Museum of Natural History, City Park, Denver 80205. Certainly an above-average study. Informs reader of the relative status of species in neighboring states when applicable. Very creditable distribution accounts. It contains no map, however, and so the reader must keep an atlas close at hand. (SRD)

Colorado Bird Distribution Latilong Study. Hugh E. Kingery and Walter D. Graul. Colorado Field Ornithologists with Colorado Division of Wildlife, 1978. 60 pages. Colorado Division of Wildlife, Nongame Section, 6060 Broadway, Denver 80216. An extraordinary synthesis of the status of 405 species recorded within the last 15 years throughout the state. The presentation employs 28 latilong blocks, shown on map on the inside cover, and incorporates field work of 83 contributors. (SRD)

Check List of Birds of Colorado. R. M. Stabler et al. 1970. 4 pages. Available from the author, Colorado College, Colorado Springs 80903.

A Pictorial Checklist of Colorado Birds, with Brief Notes on the Status of Each Species in Neighboring States of Nebraska, Kansas, New Mexico, Utah, and Wyoming. Alfred M. Bailey and Robert J. Niedrach. 1967. 168 pages. Denver Museum of Natural History, City Park, Denver 80205.

Birds of Denver and Mountain Parks. Robert J. Niedrach and Robert B. Rockwell. 1959. 207 pages. Publications Department, Denver Museum of Natural History, City Park, Denver 80205. Topography, life zones and plant associations, history, gazetteer, bibliography, index. Commentary on each species, with status, relative abundance. (JAT)

Birds of Rocky Mountain National Park. Allegra Collister. 1970. 68 pages. Denver Museum of Natural History, City Park, Denver 80205.

Birds in Western Colorado. William A. Davis. 1969. 61 pages. Historical Museum and Institute of Western Colorado, 4th and Ute, Grand Junction 81501. Bar graphs with relative abundance, habitat key, nesting information, and dates. Local areas described. Section on birding trips, special birds, and identification. (JAT)

Birder's Guide to Eastern Colorado. James A. Lane and Harold R. Holt. 1979. 126 pages. L and P Press, Box 21604, Denver 80221. Also available from ABA Sales. Introduction, bar graphs, birding trips. (JAT)

CONNECTICUT

Checklist: State of Connecticut Birds. David Junkin. 1973. 4 pages. Connecticut Audubon Society, 2325 Burr Street, Fairfield 06430.

25 Birding Areas in Connecticut. Nobel Proctor. 1978. 121 pages. Connecticut Audubon Society, Main Office, 2325 Burr Street, Fairfield 06430.

Birds of Guilford, Connecticut. Locke MacKensie. 1961. 110 pages. Publications Office, Peabody Museum of Natural History, Yale University, New Haven 06520.

A Checklist of the Birds of Southwest Fairfield County and Southeast Westchester County [New York]. Joseph Zeranski and Alice Smith. 1980. 10 pages. Mianus Naturalists' Committee of the Greenwich Audubon Society. 25¢. From Joseph Zeranski, 163 Field Pt. Rd., Greenwich, Conn. 06830. Foldout checklist on good field-stock cardboard with nesting occurrence and bar graphs that nicely show abundance through the year for 290 species.

DELAWARE

Delaware Bird List—Compiled from Published Records. John T. Linehan and Robert E. Jones. 1971. 12 pages. Society of Natural History of Delaware, J. T. Linehan, Agricultural Hall, University of Delaware, Newark 19711.

Field Check List. 1973. 4 pages. Delmarva Ornithological Society, P.O. Box 4247, Greenville 19807.

Where to Look for Birds on the Delmarva Peninsula. 1978. 42 pages. Delmarva Ornithological Society, P.O. Box 4247, Greenville 19807. A guide with maps to 37 good birding sites in the Delaware, Maryland, and Virginia sections of the Delmarva Peninsula.

Birds of Bombay Hook National Wildlife Refuge. 1972. 8 pages. Refuge Manager, Bombay Hook National Wildlife Refuge, R.D. 1, Box 147, Smyrna 19977.

Field List of the Delaware Valley Region. Alan Brady et al. 1972. 40 pages. Available from ABA Sales. Annotated field checklist, bar graphs, two maps, and general descriptions of 35 birding areas. (JAT)

FLORIDA

Florida Bird Life. Alexander Sprunt, Jr. 1954. 527 pages. Coward-McCann and the National Audubon Society. Coward-McCann, 200 Madison Avenue, New York, N.Y. 10016. An updating of the Howell-Jaques first edition. Not very much out of date. Discusses each species in detail. Jaques plates. (SRD)

Florida Bird Songs. Donald J. Borror and Maurice L. Giltz. 1980. One 12-inch, 33⅓ rpm monaural record and a 32-page manual. Dover, 180 Varick Street, New York, N.Y. 10014. Songs and calls of 59 Florida birds are presented on the record and described in the accompanying manual. Also included are line drawings of each species, a list of Florida birds, and sonagrams for each vocalization.

Where to Find Birds in Florida. M. C. Bowman and Herbert Kale. 1977. 32 pages. Available from ABA Sales.

Where to Find Birds and Enjoy Natural History in Florida. Nina Steffee and Russ Mason. 1972. 24 pages. Florida Audubon Society, P.O. Drawer 7, Maitland 32751. Detailed accounts of birding areas, with specialties, for four geographic subdivisions of the state. Many photos.

Checklist of Birds, Everglades National Park. John Ogden. 1970. 28 pages. Everglades Natural History Association, P.O. Box 279, Homestead 33030. Season, relative abundance, and comments on habitat and location in park. Map.

Check List of Birds of Palm Beach County. 1972. 4 pages. Audubon Society of the Everglades, P.O. Box 6762, West Palm Beach 33405.

Birds of the National Wildlife Refuges on the Florida Keys. 1970. 4 pages. Manager, South Florida Refuges, R.D. 1, Box 278, Delray Beach 33444.

Birds of the Loxahatchee National Wildlife Refuge. 1969. 4 pages. Manager, South Florida Refuges, R.D. 1, Box 278, Delray Beach 33444.

Birds of Sanibel-Captiva Islands, Florida. 1972. 4 pages. Sanibel-Captiva Audubon Society, Sanibel Island 33957.

Survey of the Birdlife of Northwestern Florida. Bulletin 5 of Tall Timbers Research Station, 1965. 147 pages. Tall Timbers Research Station, Route 1, Box 160, Tallahassee 32312. Available through Xerox University Microfilms, OP Book Department, 300 N. Zeeb Road, Ann Arbor, Mich. 48106. Description of region, including climate. Thorough discussion of migration through area and recent changes affecting bird life. Ornithological history of area. Very detailed species accounts. References. (JAT)

GEORGIA

Georgia Birds. Thomas D. Burleigh. 1958. 746 pages. University of Oklahoma Press, 1005 Asp Avenue, Norman, Okla. 73019. An acceptable though not thorough account. (SRD)

Annotated Checklist of Georgia Birds. J. F. Denton et al. 1977. 66 pages. Georgia Ornithological Society, P.O. Box 38214, Atlanta 30334.

Birder's Guide to Georgia. D. W. Hans, ed. Georgia Ornithological Society. 1976. Out of print, new edition in preparation.

Birds of Okefenokee. 1969. 6 pages. Manager, Okefenokee National Wildlife Refuge, Box 117, Waycross 31501.

Field Checklist of West-Central Georgia Birds. Compiled by Fred Galle. 1965. 4 pages. Available free with self-addressed, stamped envelope from Ida Cason, Calloway Gardens, Pine Mountain 31822.

HAWAII

Hawaiian Birdlife. Andrew J. Berger. 1973. 270 pages. University Press of Hawaii, 2040 Kolawalu Street, Honolulu 96822. A thorough, well-organized book which can be considered the definitive work on the avifauna of Hawaii. It postdates

Munro's *Birds of Hawaii.* Detailed discussions on the history of the islands and the variety of habitats that evolved through approximately 15–20 million years of isolation. (SRD)

Hawaii's Birds. Robert J. Schollenberger, ed. 1978. 100 pages. Hawaii Audubon Society, Box 22832, Honolulu 96822. Seventy-one color photos. Guide to Hawaiian birds including 47 endemics.

Preliminary List of the Birds of Hawaii. Robert L. Pyle. 1977. 12 pages. Hawaii Audubon Society, Box 22832, Honolulu 96822.

Guide to Hawaiian Birding. C. J. Ralph, ed. 1977. 10 pages. Hawaii Audubon Society, P.O. Box 22932, Honolulu 96822.

Birds of the Hawaiian Islands National Wildlife Refuge. 1971. 6 pages. Manager, Hawaiian Islands National Wildlife Refuge, 337 Uluniu Street, Kailua 96734.

Preliminary Checklist of the Birds of Hawaii, Volcanoes National Park. 1965. 1 page. Superintendent, Hawaii Volcanoes National Park 96718.

IDAHO

Birds of Idaho. Thomas D. Burleigh. 1972. 467 pages. Caxton Printers Ltd., P.O. Box 700, Caldwell 83605. The first detailed work on the birds of Idaho. Each species writeup lists records of Burleigh and others, species range and status in Idaho, a short section on species' habitats. Bibliography. Photos and map leave much to be desired. (SRD)

Checklist of the Birds of Idaho. 1969. 4 pages. Migration Information, Idaho Fish and Game Department, 600 S. Walnut Street, P.O. Box 25, Boise 83707.

Birds of Southeastern Washington. John W. Weber and Earl J. Lanison. 1979. 66 pages. University Press of Idaho, P.O. Box 3368, University Station, Moscow 83843. This distribution checklist includes discussion of recent changes in bird populations as well as notes on geography and climatology of southeastern Washington and neighboring Idaho.

Birds of Craters of the Moon National Monument. D. L. Carter. 1970. 2 pages. Superintendent, Craters of the Moon National Monument, Arco 83213.

Birds of the Deer Flat National Wildlife Refuge. 1969. 6 pages. Manager, Deer Flat National Wildlife Refuge, Route 1, Box 1457, Nampa 83651.

Birds of the Kootenai National Wildlife Refuge. 1972. 8 pages. Manager, Kootenai National Wildlife Refuge, Box 88, Star Route 1, Bonners Ferry 83805.

ILLINOIS

Field Checklist of Illinois Birds. 1973. 4 pages. Illinois State Museum, Spring and Edwards Streets, Springfield 62706.

Bird Finding in Illinois. Fawks and Lobik. 1975. 90 pages. Illinois Audubon Society, 34 West, 269 Whitethorn Road, Wayne 60184.

Chicagoland Birds. Smith and Beecher. 1972. 57 pages. Bookstore, Field Museum of Natural History, Lake Shore Drive and Roosevelt Road, Chicago 60605. Birding calendar, major birding areas, bird clubs, and bar graphs with habitat and locality keys. Three maps and space for 12 days' records.

Top Birding Spots Near Chicago. Jeffrey Sanders and Lynn Yaskot. 1975. 48 pages. Available from ABA Sales. 15 maps and guide to 36 top birding areas around Chicago. (JAT)

Field List of the Birds of Southern Illinois. Vernon Kleen. 1971. 20 pages. Vernon Kleen, Natural Heritage Section, Illinois Department of Conservation, 605 Stratton Building, Springfield 62706.

Birding Handbook to East-Central Illinois. John Behrens, ed. 1976. 56 pages. Available from ABA Sales. A guide to some of the best birding areas. Annotated checklist. (JAT)

INDIANA

Field Checklist of Indiana Birds. 1978. 4 pages. Indiana Audubon Society, Inc., Mary Gray Bird Sanctuary, Route 6, Connersville 47331.

Annotated Checklist of Indiana Birds. Russell Mumford and Charles Keller. 1975. 63 pages. Indiana Audubon Society, Inc., Mary Gray Bird Sanctuary, Route 6, Connersville 47331. A complete list of Indiana's birds with bar graphs and annotations.

IOWA

Birding Areas of Iowa. Edited by Peter C. Petersen. 1979. 152 pages. Available from Iowa Ornithologists' Union, c/o Mrs. Pat Layton, 1560 Linmar Drive, Cedar Rapids 52404.

Iowa Ornithologists' Union Field Check List. 1974. 4 pages. Mrs. Ruth Buckles, 5612 Urbandale Avenue, Des Moines 50310.

Birds, DeSoto National Wildlife Refuge. 1968. 6 pages. Manager, DeSoto National Wildlife Refuge, R.R. 1B, Missouri Valley 51555.

Field List of Birds of the Quad-City Region. 1977. 27 pages. Available from ABA Sales. Annotated list of the birds occurring in the vicinity of Davenport. Brief descriptions of 31 birding areas. (JAT)

KANSAS

Directory to the Birds of Kansas. Richard F. Johnston. 1965. 67 pages. University of Kansas, Museum of Natural History, Lawrence 66044. A modest publication with limited coverage of abundance, habitat preference, dates of occurrence, breeding schedule, clutch size and additional literature citations for the 379 species (461 species and subspecies) known to have occurred in Kansas. (SRD)

Birds in Kansas. Arthur L. Goodrich. 1945. 340 pages. State Board of Agriculture, 503 Kansas Avenue, 4th floor, Topeka 66603. A traditional state bird book composed of all the usual features with one unique component: an explanation of the scientific name of each species. The line drawings are commendable. (SRD)

Hand-List of the Birds of Kansas. Richard F. Johnston. Miscellaneous Publication No. 22, 1960. 6 pages. University of Kansas, Museum of Natural History, Lawrence 66044.

Birds of the Cheyenne Bottoms Waterfowl Management Area. 1974. 8 pages. Barton County Community College, Great Bend 67530.

KENTUCKY

Kentucky Birds. R. W. Barbour et al. 1975. University of Kentucky Press, Lexington 40506. Excellent coverage of the state birds including abundance, seasonal occurrence, habitat preference, and breeding details. The last third of the book is concisely devoted to bird-finding in the state. (SRD)

Birds of Kentucky. Robert M. Mengel. The American Ornithologists' Union. Ornithological Monographs No. 3, 1965. 581 pages. National Museum of Natural History, Smithsonian Institution, Washington, D.C. 20560. A veritable object lesson in the state bird book as we know it. Covers habitat, range status, abundance, and migration. Contains model discussions on those species whose range is expanding or should be expanding and also superb range maps. (SRD)

Summary of Occurrence of Birds of Kentucky. Monroe. 1969. 10 pages. Mrs. Clifford F. Johnson, 1166 Castlevale, No. 4, Louisville 40217. Introduction, bar graphs, nesting status. Indicates occurrence in 4 areas of state.

Checklist of Kentucky Birds. Kentucky Ornithological Society. 4 pages. Available free with self-addressed, stamped envelope from Mrs. Clifford F. Johnson, 1166 Castlevale, No. 4, Louisville 40217.

LOUISIANA

Louisiana Birds, 3rd revised edition. George H. Lowery, Jr. 1974. 651 pages. Louisiana State University Press, Baton Rouge 70803. A completely revised state bird book, with updated seasonal occurrence charts, as well as thorough and well-researched text. (SRD)

Field Checklist of Louisiana Birds. Available for 5¢ and a self-addressed, stamped envelope from Louisiana Ornithological Society, Museum of Natural Sciences, Louisiana State University, Baton Rouge 70803.

Birds of the Delta National Wildlife Refuge. 1967. 4 pages. Manager, Delta–Gulf Islands National Wildlife Refuge, 1216 Amelia Street, Gretna 70053.

Birds of Lacassine National Wildlife Refuge. 1964. 4 pages. Manager, Lacassine National Wildlife Refuge, Route 1, Box 186, Lake Arthur 70549.

Birds of the Sabine National Wildlife Refuge. 1963. 4 pages. Manager, Sabine National Wildlife Refuge, MRH 107, Sulphur 70645.

MAINE

Maine Birds. Ralph S. Palmer. Bulletin of the Museum of Comparative Zoology at Harvard College, Vol. 102, 1949. 656 pages. Museum of Comparative Zoology, Harvard College, Cambridge, Mass. 02139. Plan of this thorough book includes spring and fall migration information with extreme dates and, where possible, peak dates, species' status summary, notes on flight years or incursions, breeding and/or summer data, winter status, ecological and general remarks on historical changes in status, behavior, taxonomy. Discusses more than 365 species and contains an extensive citation listing. Needs updating. Excellent and recommended. (SRD)

Annotated Checklist of Maine Birds. P. D. Vickery. 1978. 20 pages. Maine Audubon Society, 118 Route One, Gilsland Farm, Falmouth 04105. State list with abundance categories, graphs of occurrence patterns, and preferred habitats.

Enjoying Maine Birds. O. S. Pettingill, Jr., ed; revised by R. B. Anderson and I. Richardson. 1972. 84 pages. Maine Audubon Society, 118 Route One, Gilsland Farm, Falmouth 04105. Contains line illustrations and species accounts of 80 common Maine birds, discussion of bird-finding sites in Maine, checklist to Maine birds as well as comments on bird migration, research, photography, and methods to attract birds.

Native Birds of Mt. Desert Island. James Bond. 1969. 28 pages. Academy of Natural Sciences of Philadelphia, 19th and The Parkway, Philadelphia, Pa. 19103. Species accounts for 138 species occurring from June to August; all have bred in last 50 years. (JAT)

Checklist of the Birds of Acadia National Park. Paul Favour, Jr. 1969. 15 pages. Superintendent, Acadia National Park, Hulls Cove 04644.

Birds of Moosehorn Wildlife Refuge. 1970. 6 pages. Manager, Moosehorn National Wildlife Refuge, Box X, Calais 04619.

MARYLAND/WASHINGTON, D.C.

Birds of Maryland and the District of Columbia. Robert E. Stewart and Chandler S. Robbins. North American Fauna No. 62, U.S. Department of the Interior, 1958. 401 pages. U.S. Government Printing Office, Division of Public Documents, Washington, D.C. 20402. Covers all of the usual material as well as analyzing banding recovery data, maximum counts, extreme dates, comparative abundance, maps of biotic areas, and sections in Maryland where each species may be expected and when. Out of print and out of date with respect to more obscure species (e.g., pelagics) and predates several new additions (e.g., cattle egret, house finch, American avocet). (SRD)

Field List of the Birds of Maryland, 2nd edition. Chandler S. Robbins and Danny Bystrak. 1977. 52 pages. Maryland Avifauna No. 2. Maryland Ornithological Society, Clyburn Mansion, 4915 Greenspring Avenue, Baltimore 21209. Also available from ABA Sales. Checklist with bar graphs, directions, and map to 50 choice birding areas.

Bird Checklist of National Capital Parks. 4 pages. National Parks Service, Washington, D.C.

Birds Recorded on Patuxent Wildlife Research Center, 1941–1969. Chandler S. Robbins. 1970. Migratory Bird Population Station, Laurel 20810.

Birds of the Blackwater National Wildlife Refuge. 1970. 4 pages. Manager, Blackwater National Wildlife Refuge, Box 121, Route 1, Cambridge 21613.

Birds of Eastern Neck National Wildlife Refuge. 1971. 8 pages. Manager, Eastern Neck National Wildlife Refuge, Route 2, Box 225, Rock Hall 21661.

MASSACHUSETTS

Birds of Massachusetts and Other New England States. Edward Howe Forbush. 1925–29. Massachusetts Department of Agriculture, 100 Cambridge Street, Boston 02202. The Louis Agassiz Fuertes plates alone raise this set far above the average.

The information is superior and not altogether outdated. It is an indispensable classic. (SRD)

Birds of Massachusetts: An Annotated and Revised Check List. Ludlow Griscom and Dorothy E. Snyder. 1955. 295 pages. Peabody Museum of Salem, East India Square, Salem 01970. Dated, but remains much more than an annotated checklist. (SRD)

List of the Birds of Massachusetts. Bradford Blodget. 1979. 20 pages. Massachusetts Division of Fisheries and Wildlife, Leverett Saltonstall Building, Government Center, 100 Cambridge Street, Boston 02202.

Checklist of Massachusetts Birds. 1974. 4 pages. Massachusetts Audubon Society, Lincoln 01773.

Where to Find Birds in Eastern Massachusetts. L. J. Robinson and R. A. Stymeist, eds. 1978. 176 pages. *Bird Observer,* 462 Trapelo Road, Belmont 02178.

Bird-Finding in Massachusetts. Massachusetts Audubon Society, Lincoln 01773.

Birds of the Cape Cod National Seashore and Adjacent Areas. Wallace Bailey. 1968 (with 1970 supplement). 134 pages. Massachusetts Audubon Society, Lincoln 01773. Detailed species accounts, including hypotheticals. Line drawings and descriptions of 50 common birds. Eighteen birding areas described. (JAT)

Birds of Martha's Vineyard with an Annotated Checklist. L. Griscom and E. V. Folger. 1959. 164 pages. Massachusetts Audubon Society, Lincoln 01773.

MICHIGAN

Birds of Michigan. Norman W. Wood. Miscellaneous Publications No. 75, Museum of Zoology, University of Michigan, 1951. 559 pages. University of Michigan, Museum of Zoology, 1109 Geddes Road, Ann Arbor 48109. Fully documented account of the status of Michigan birds. Limited to distribution, relative abundance, and migration. No color plates, plumage descriptions, or detailed accounts of habits or habitats. Contains one Michigan counties map, 16 black-and-white plates of poor quality and dubious merit, and a reasonably fat bibliography. (SRD)

A Checklist of Michigan Birds. Victor S. Janson and Dr. Lawrence A. Ryel. 1980. 10 pages. Information Services Center, Michigan Department of Natural Resources, Box 30028, Lansing 48909.

Birds of S.E. Michigan and S.W. Ontario. Alice H. Kelly. 1978. 99 pages. Available from ABA Sales. Detailed species accounts of over 337 species from this region. (JAT)

Enjoying Birds in Michigan. William L. Thompson. 1970. 80 pages. Michigan Audubon Society, 7000 N. Westnedge Avenue, Kalamazoo 49007. Many chapters on a variety of topics relating to birds and birding. Fourteen pages on distribution and bird finding, including a checklist with symbols for breeding, section of state, and rare or accidental species. Maps of 6 sanctuaries and pictorial map of state showing general location of 53 birding areas. (JAT)

Guide to Bird-Finding in Washtenaw County and Surrounding Areas. Edited by Alfred

Maley. 1971. 16 pages. Detroit Audubon Society, 814 W. Seven Mile Road, Detroit 48236. Detailed directions to favored birding areas in southeastern Michigan, northwestern Ohio, southern Ontario, and northern Michigan. (JAT)

Birds of Isle Royale in Lake Superior. Laurits W. Krefting et al. Fish and Wildlife Service, U.S. Department of the Interior, Special Scientific Report—Wildlife No. 94, 1966. 56 pages. Superintendent, Isle Royale National Park, Houghton 49931. Describes habitats, has checklist arranged by seasonal abundance and habitats. Annotated list of birds. References and plant list. (JAT)

Seasonal Distribution and Abundance of Birds in the Kalamazoo, Michigan, Area. Raymond J. Adams, Jr. 1979. 16 pages from the *Jack Pine Warbler.* Kalamazoo Nature Center, 7000 N. Westnedge Avenue, Kalamazoo 49007. A calendar list showing abundance and dates of occurrence based on 150 years of bird observation in the Kalamazoo area.

MINNESOTA

Minnesota Birds: When and How Many. Janet Green and Robert B. Janssen. 1975. 217 pages. University of Minnesota Press, 2037 University Avenue, S.E., Minneapolis 55455. Authoritative, impressive, exemplary state bird book treating the migration, distribution, and seasonal status of the 374 species recorded in Minnesota. Species accounts are lucid and distribution maps well executed. (SRD)

Manual for the Identification of the Birds of Minnesota and Neighboring States, revised edition. Thomas Sadler. 1955. 279 pages. University of Minnesota Press, 2037 University Avenue, S.E., Minneapolis 55455. Taken from the famous Roberts's *Birds of Minnesota,* this key still stands as an advanced-level reference for those who have the bird in hand and require more information about fine points of identification, distribution, age determination, and migratory status than the bare bones offered in most field guides. (SRD)

Daily Field Checklist of Minnesota Birds. Minnesota Ornithologists Union, 1974. 4 pages. Bookstore, Bell Museum of Natural History, 10 Church Street, S.E., University of Minnesota, Minneapolis 55455.

A Birder's Guide to Minnesota. Kim Eckert. Minnesota Ornithologists Union, 1974. 114 pages. Bookstore, Bell Museum of Natural History, 10 Church Street, S.E., University of Minnesota, Minneapolis 55455. Introduction, annotated list of Minnesota birds, 150 good birding areas described, 6 maps, and daily field card included. (JAT)

Birds of the Minneapolis–St. Paul Region. Anne Dodge et al. 1966. 30 pages. Bookstore, Bell Museum of Natural History, 10 Church Street, S.E., University of Minnesota, Minneapolis 55455. Bar graphs, habitat key, and space for 8 days' records.

MISSISSIPPI

Bird Life of the Gulf Coast of Mississippi. Thomas D. Burleigh. Louisiana State University Museum of Zoology Occasional Paper No. 20, 1944. 166 pages. Louisiana State Museum of Zoology, Baton Rouge, La. 70803. Does not cover the entire state of Mississippi but is an excellent reference for the birds of the Gulf Coast. Predictably dated but not antique. (SRD)

Field Checklist. 1974. 4 pages. Mississippi Ornithological Society, c/o Museum of Natural Science, 111 N. Jefferson Street, Jackson 39202.

MISSOURI

Checklist of Missouri Birds. David Easterla and Richard Anderson. 1971. 4 pages. Mrs. Edna Kriege, 6123 Waterman, St. Louis 63112.

Time Table for Birds of Kansas City. Shirling et al. 1967. 8 pages. Rilla Hammat, 4829 Holmes, Kansas City 64110.

Guide to Finding Birds in the St. Louis Area. Anderson and Bauer. 1968. 44 pages. Webster Groves Nature Study Society, St. Louis County Library, Clayton and Lindbergh, St. Louis 63131. Preface and introduction. Twenty-one pages of descriptions and maps of birding areas. Bar graphs and comments on distribution. (JAT)

Introduction to Bird Study in Missouri. David Easterla and Richard Anderson. 1979. 18 pages. Checklist available free from Missouri Department of Conservation, P.O. Box 180, Jefferson City 65102.

MONTANA

Montana Bird Distribution. P. D. Skaar. 1975. 56 pages. Available from the author, 501 S. Third, Bozeman 59715. The author has chosen a unique scheme to illustrate the distributional status of every one of the 357 species: maps based on latilongs, which are areas between lines of latitude and longitude. The scheme is simple and should be eminently useful. (SRD)

Birds of Montana. C. V. Davis. 1973. 6 pages. Available from the author, 1620 S. Third, Bozeman 59715.

Birds of the Bozeman Latilong. P. D. Skaar. 1969. 132 pages. Available from the author, 501 S. Third, Bozeman 59715. Detailed description of area covered (which includes parts of Idaho, Wyoming, and Yellowstone National Park), and of several birding "hot spots." History of coverage. Very detailed species accounts provide a wealth of information on abundance and dates. Bibliography and index. (JAT)

NEBRASKA

Revised Checklist of Nebraska Birds (with Supplement through 1970). Rapp et al. 48 pages. Librarian, Nebraska Ornithologists' Union, University of Nebraska State Museum, Lincoln 68508. Introduction includes map, description of zoogeography. Brief species accounts, bibliography, index. (JAT)

Daily Field Record of Nebraska Birds. 8 pages. Librarian, Nebraska Ornithologists' Union, University of Nebraska State Museum, Lincoln 68508.

NEVADA

Birds of Nevada. Jean M. Linsdale. Pacific Coast Avifauna No. 23, 1936. Supplement published in *The Condor,* Vol. 53, No. 5, 1951. Cooper Ornithological Society, c/o Department of Biology, UCLA, 405 Hilgard Avenue, Los Angeles, Calif. 90024. Too old to be of much value. The supplement that appeared in *The Condor* renders the 1936 Linsdale at least usable, but Nevada's birds could benefit from a fresh approach and a thoroughgoing revision. (SRD)

Checklist of the Birds of Nevada. Fred Ryser. 1970. 12 pages. Available free with self-addressed, stamped envelope from the author, Biology Department, University of Nevada, Reno 89507.

Checklist of the Birds of Southern Nevada. C. S. Lawson. 1977. 20 pages. Available from the author, 509 Altamira Road, Las Vegas 89128, and ABA Sales. A complete list of Nevada's birds in a ten-column field-checklist format. (JAT)

NEW HAMPSHIRE

Checklist of Birds in New Hampshire. Tudor Richards. 1964. 12 pages. Audubon Society of New Hampshire, 3 Silk Farm Road, Concord 03301. Dates of occurrence and favorite habitats for state. Status, relative abundance for each of 3 divisions of state. (JAT)

Birding in Northern New Hampshire. Vera H. Hebert. 1965. 6 pages. Available free with self-addressed, stamped envelope from Audubon Society of New Hampshire, 3 Silk Farm Road, Concord 03301. List and birding area directions for the Connecticut Lakes region and the Androscoggin Valley region. (JAT)

Monadnock Sightings: Birds of Dublin, New Hampshire 1909–1979. Elliott and Kathleen Allison, introduction by Edwin Way Teale. 1979. 55 pages. William L. Bauhan, Dublin, New Hampshire 03444. An annotated checklist that compares the authors' observations of Dublin birds with those of Gerald Thayer published seventy years earlier. A useful guide for Mt. Monadnock.

NEW JERSEY

Birds of New Jersey: Their Habits and Habitats. Charles Leck. 1975. 190 pages. Rutgers University Press, 30 College Avenue, New Brunswick 08903. This state book does not follow traditional lines but rather discusses major avian communities: the bird life of well-defined habitats. The maps and habitat/bird discussions are less precise than is desirable, but neophyte birders in New Jersey will count it among their aids in bird finding. (SRD)

Bird Studies at Old Cape May. Witmer Stone. 1937. 2 vols. 941 pages. Dover Publishers, 180 Varick Street, New York, N.Y. 10014. Monumental landmark study of the birds of Cape May but not limited to that area alone. These volumes cover the entire coast of New Jersey, with emphasis on the southern half of the state. A classic regional work. (SRD)

Checklist for Birds of New Jersey. 1964. 40 pages. Public Information Office, New Jersey Department of Conservation and Economic Development, Trenton 08625.

Checklist for the Birds of the Sandy Hook Area. Mario Digregario and Bill Smith. 1978. 5 pages. Sandy Hook Visitor Center, Gateway National Recreation Area, Sandy Hook Unit, Box 437, Highlands 07732.

New Jersey Field Trip Guide. William J. Boyle, Jr. Summit Nature Club, 1979. 55 pages. Available from William J. Boyle, Jr., 15 Indian Rock Road, Warren 07060. A map, general directions, specific trail directions, and matters of interest are given for each of 23 birding areas in New Jersey, for Jamaica Bay Refuge in New York, and for Hawk Mountain, Pa.

Birds of Great Swamp National Wildlife Refuge. 1972. 8 pages. Manager, Great Swamp National Wildlife Refuge, R.D. 1, Box 148, Basking Ridge 07920.

Birds of Brigantine. 1977. 10 pages. Manager, Brigantine National Wildlife Reserve, Great Creek Road, P.O. Box 72, Oceanville 08231.

NEW MEXICO

Revised Check-list of the Birds of New Mexico. John P. Hubbard. New Mexico Ornithological Society Publication No. 6, 1978. 110 pages. John P. Hubbard, 2016 Valle Rio, Santa Fe 87501. Completely updates the state's avifauna following the format used in the 1970 edition, addressing frequency of occurrence, status, range, numbers, habitats, and elevations for each of the 433 species confirmed in the state. (SRD)

New Mexico Birds. Stokley J. Ligon. 1961. 360 pages. University of New Mexico Press, Albuquerque 87131. Fairly complete treatment of the birds of New Mexico but is becoming outdated quickly. This work lacks details such as peak migration times, biggest counts, precise and distinguished maps. A concentrated modernization would enhance the present work. (SRD)

Field Card. New Mexico Ornithological Society. 4 pages. William Baltosser, Department of Biology, New Mexico State University, Las Cruces 88003.

Sand and Feathers: Birds of White Sands National Monument. Compiled by George T. Morrison, 1975. Revised by Dennis Vasquez, 1979. 6 pages. Superintendent, White Sands National Monument, P.O. Box 458, Alamogordo 88310. Field checklist for the 190 species observed over a 40-year period within the boundaries of the national monument.

Bird Species of the Organ Mountains Wildlife Habitat Area. 6 pages. District Manager, U.S. Department of the Interior, Bureau of Land Management, Box 1420, Las Cruces 88001.

Packet of "Good Birding Trips." 1975. 5 pages. Central New Mexico Audubon Society, P.O. Box 30002, Albuquerque 87190. Packet of 5 field cards listing notable species with maps showing travel directons. To be revised in 1981.

Summer Birds of San Juan Valley, New Mexico. C. Gregory Schmitt. 1976. 22 pages. New Mexico Ornithological Society Publication No. 4. John P. Hubbard, 2016 Valle Rio, Santa Fe 97501.

Breeding Birds of Elephant Butte Marsh. Charles A. Hundertmark. 1978. 17 pages. New Mexico Ornithological Society Publication No. 5. John P. Hubbard, 2016 Valle Rio, Santa Fe 97501.

NEW YORK

Birds of New York State. John Bull. 1974. 655 pages. Doubleday and Company, 501 Franklin Avenue, Garden City 11530. Without question an essential reference for the student of New York State avifauna; includes in Part 2—Family and Species Accounts—discussion of 410 species. Voluminous data on the geographical distribution, range changes, seasonal occurrence, nesting period, egg dates, and bird banding recoveries analysis. Included are 86 black-and-white photographs and nine color plates. There is an index, gazetteer, and bibliography. The 1976 Supplement (see below) brings the records up to date. (SRD)

Supplement to Birds of New York State. John Bull. 1976. 52 pages. Federation of New York State Bird Clubs (FNYSBC), 533 Chestnut Street, West Hempstead 11552. This supplement brings up to date (to July 1, 1975) Bull's *Birds of New York State* (see above). There are updated species accounts, references, and corrigenda listings. Two new species have been recorded in the state, and another, elevated from "hypothetical" to occurrence status, brings the state list to 413 species. An addition of 3 new breeding species brings that total to 231. It is intended that annual supplements will appear in *The Kingbird,* the quarterly publication of FNYSBC. (SRD)

Birds of the New York Area. John Bull. 1964. 540 pages. Dover Publishers, 180 Varick Street, New York 10014. This is a thorough treatment of special value to the residents of the New York City area. Detailed accounts of the 431 species known to occur within 35 miles of Manhattan.

Birdlife of the Adirondack Park. Bruce M. Beehler. 1978. 210 pages. Adirondack Mountain Club, Inc., 172 Ridge St., Glens Falls 12801. Also available from ABA Sales. Habits, distribution, and abundance of Adirondack birds. Illustrations and photos.

Birds of Essex County, New York. Geoffrey Carleton. 1980. 35 pages. High Peaks Audubon Society, Inc., Discovery Farm, R.D. 1, Elizabethtown 12932. A guide to 270 bird species and 20 birding areas in the Adirondack-Champlain region.

Birds of Wyoming County, New York. Richard C. Rosche. Bulletin of the Buffalo Society of Natural Sciences, vol. 23, 1967. 89 pages. Buffalo Museum of Science, Buffalo 14211. Introduction, description of area, and description of 12 birding areas. Very detailed species accounts. Ten maps, 9 of which show the summer distribution of 9 species. Checklist with status, breeding status, and migration dates. (JAT)

Bird-finding in Onondaga County. Dorothy Crumb. 1971. 23 pages. Onondaga Audubon Society, P.O. Box 620, Syracuse 13212. Bird-finding directions for 20 towns.

Enjoying Birds Around New York City. Robert Arbib, Olin Sewall Pettingill, and Sally H. Spofford. 1966. 162 pages. Cornell Laboratory of Ornithology, 159 Sapsucker Woods Road, Ithaca 14850.

Birding in the Cayuga Lake Basin. Mildred Comar, Douglas Kibbe, and Dorothy McIlroy. 1974. 108 pages. Cornell Laboratory of Ornithology, 159 Sapsucker Woods Road, Ithaca 14850. Maps.

Enjoying Birds in Upstate New York. Olin S. Pettingill and Sally H. Spofford. 1975. 91 pages. Cornell Laboratory of Ornithology, 159 Sapsucker Woods Road, Ithaca 14850.

Migratory Chronology and Checklist of Western New York State Birds. James W. Keefer.

1979. 32 pages. Available from the author, 244 Hill Top Lane, Spencerport 14559. This booklet includes a checklist for western New York State birds (east to Tompkins County), habitat preferences for 358 species, breeding status, and migration charts that show seasonal occurrences for all species.

NORTH CAROLINA

Birds of North Carolina, revised edition. Thomas G. Pearson, Clement Samuel Brimley, and Herbert Hutchinson Brimley. 1959. 434 pages. North Carolina Department of Agriculture, P.O. Box 27647, Raleigh 27611. The updated version of the 1942 edition. It gives considerable documentation and attention to the new species added to the state list and totals it at 408 species. There are no distribution maps, no simple relative abundance graphic representations, no interesting color or black-and-white plates. (SRD)

Birds of the Carolinas. Eloise F. Potter et al. 1980. 408 pages. The University of North Carolina Press, Box 2288, Chapel Hill 27514. More than 400 species accounts describe the birds of North and South Carolina. The accounts are illustrated by 338 color photographs.

Checklist of Birds for Cape Hatteras National Seashore. 6 pages. Superintendent, Cape Hatteras National Seashore, Box 457, Manteo 27954.

Birds of Mackay Island National Wildlife Refuge. 1971. 4 pages. Manager, Back Bay National Wildlife Refuge, Box 6128, Virginia Beach, Va. 23456.

Birds of the Pee Dee National Wildlife Refuge. 1971. 4 pages. Manager, Pee Dee National Wildlife Refuge, Box 780, Wadesboro 28170.

NORTH DAKOTA

Breeding Birds of North Dakota. Robert E. Stewart. 1975. 295 pages. Tri-College Center for Environmental Studies, Fargo 58102. Limited to past and present status of northern prairie and interior plains breeding birds. Excellent species accounts comprehensively treat breeding habitat, range (with many and extensive maps), and nesting. (SRD)

Field Checklist—Birds of the Bismarck–Mandan Area, N.D. Bismarck-Mandan Bird Club. 1978. 8 pages. 25¢. From Robert N. Randall, 928 N. 16th St., Bismarck 58501. Foldout field card listing species and seasonal abundance found in the vicinity of southcentral North Dakota.

A Birder's Guide to North Dakota. Kevin Zimmer. 1976. 114 pages. L and P Press, Box 21604, Denver, Colo. 80221. Over 40 birding localities described in detail.

Birds of the Grasslands. 1973. 16 pages. Theodore Roosevelt Nature and History Association, Medora 58645. Checklist.

OHIO

Birds of Buckeye Lake. Milton B. Trautman. Museum of Zoology Miscellaneous Publication No. 44, 1940. 466 pages. University of Michigan, Museum of Zoology, 1109 Geddes Road, Ann Arbor, Mich. 48109. A systematic account of 282 species and six subspecies recorded in the Buckeye Lake area with extensive analysis and historical account of the vegetation and avifauna therein. (SRD)

Annotated List of the Birds of Ohio. Milton B. Trautman and Mary A. Trautman. 1968. 77 pages. Ohio Academy of Science, 445 King Avenue, Columbus 43201. Complete list, including migration dates, status and abundance, extensive bibliography.

Checklist of the Birds of Ohio. T. Thomson. 1974. 8 pages. Columbus Audubon Society, 1065 Kendale Road N., Columbus 43220.

Birding in Ohio. Tom Thomson. In preparation by author, 404 Thurber Drive W., Columbus 43215. One hundred forty good birding sites are reviewed, with travel directions, bits of historical and geological background, notes on trees, flowers, and ferns, and birds that are apt to be seen during various seasons. Also included in the book are the following features: a complete checklist of the birds of Ohio, a list of birding organizations, changes in bird population since 1950, and a migration calendar.

Birds of the Cleveland Region. Donald L. Newman. 1969. 48 pages. Cleveland Museum of Natural History, Wade Oval, University Circle, Cleveland 44106. Introduction, ornithological history, several pages of noteworthy records, and explanatory comments on selected species. Directions to 18 favored birding areas, with 45 additional ones indicated on map of region (30-mile radius from Cleveland public square including Akron lakes). Bar graphs and nesting indication. Index. (JAT)

Birds of the Toledo Area. Louis W. Campbell. 1968. 330 pages. *The Toledo Blade,* Promotion and Public Service, Orange and Superior Streets, Toledo 43660. Sections on physical geography of area, migration lanes, ornithological history, and seasonal changes in bird life. Very detailed species accounts. (JAT)

Birds of the Ottawa, Cedar Point, and West Sister Island National Wildlife Refuges. 1970. 4 pages. Manager, Ottawa National Wildlife Refuge, R.R. #3, Box 269, Oak Harbor 43449.

The Birds of Hueston Woods State Park. 8 pages. Department of Natural Resources, Division of Parks and Recreation, Fountain Square, Columbus 43224.

OKLAHOMA

Oklahoma Birds. George Miksch Sutton. 1967. 674 pages. University of Oklahoma Press, 1005 Asp Avenue, Norman 73019. Data has been gathered from all of the 77 counties, from all elevations, during all seasons, night and day, with special attention paid to 11 physiographic regions, 6 vegetational regions, 10 biotic districts, and 12 "game"-type areas—and once gathered has been interestingly discussed and charmingly illustrated with 29 of Sutton's own works. (SRD)

Checklist of Birds of Oklahoma. Gary D. Schnell. 1978. 4 pages. Stovall Museum, 1335 Asp Avenue, Norman 73019.

Birds of Southwestern Oklahoma. Jack D. Tyler. 1979. 65 pages. Contribution No. 2, Stovall Museum, University of Oklahoma, Norman 73019.

A Supplement to Birds of Southwestern Oklahoma. Jack D. Tyler. 1979. 55 pages. Contribution No. 3, Stovall Museum, University of Oklahoma, Norman 73019.

Checklist of the Birds of Platt National Park. 10 pages. Superintendent, Platt National Park, P.O. Box 201, Sulphur 73086.

Birds of the Salt Plains National Wildlife Refuge. 1968. 4 pages. Manager, Salt Plains National Wildlife Refuge, Jet 73749.

Tulsa Audubon Society Bird Finding Guide. 1973. 118 pages. Tulsa Audubon Society, Central Library, 400 Civic Center, Tulsa 74100. Complete information for birding near Tulsa, including roads, habitats, and species to be expected. Thirty-six maps.

OREGON

Birds of Oregon. Ira N. Gabrielson and Stanley G. Jewett. 1940. 650 pages. Revised edition issued as *Birds of the Pacific Northwest,* 1970. Dover Publishers, 180 Varick Street, New York, N.Y. 10014. An excellent reference work in spite of its vague and crude distribution maps (which are often inscrutable). Authors are heavy on subspecies maps that are, for the most part, deficient. Otherwise a superior work. (SRD)

Birding Oregon. Fred L. Ramsey. 1978. 178 pages. Available from ABA Sales. Oregon's first birding guide contains maps and directions for prime birding spots in seven different habitats. (JAT)

Birds of Mt. Hood's South Slope. 8 pages. U.S. Forest Service, 319 Pine Street, P.O. Box 3623, Portland 97208.

Birds of the Cold Springs National Wildlife Refuge. 1968. 4 pages. Manager, Umatilla National Wildlife Refuge, Box 239, Umatilla 97882.

Birds of the Malheur National Wildlife Refuge. 1978. 8 pages. Manager, Malheur National Wildlife Refuge, Box 113, Burns 97720.

PENNSYLVANIA

Birds of Western Pennsylvania. Walter Edmond Clyde Todd. 1940. 710 pages. University of Pittsburgh Press, 127 N. Bellefield Avenue, Pittsburgh 15260. Covers only the western half of the state. Maps indicating breeding locales are remarkably modern. Outshines most traditional state works by remaining timely. (SRD)

Pennsylvania Birds: An Annotated List. Earl L. Poole. 1964. 104 pages. Delaware Valley Ornithological Club, Academy of Natural Sciences of Philadelphia, 19th and The Parkway, Philadelphia 19103. Ornithological history, physiography, maps. Very detailed species accounts, bibliography. (JAT)

Birds of Pennsylvania. Merrill Wood. 1979. 148 pages. The Pennsylvania State University, Box 6000, University Park 16802. Discusses topography, climate, migration seasons. Bird-life calendar. Species accounts, many line drawings. References and index. (JAT)

Birding the Delaware Valley Region. John J. Harding and Justin J. Harding. 1980. 176 pages. Temple University Press, Broad and Oxford Streets, Philadelphia 19122. Here are detailed directions and maps to seventy birding sites within a two-hour drive of Philadelphia. Annotated discussion of 335 species and where and when they can best be observed.

Field List of Birds of the Pittsburgh Region. Kenneth C. Parkes. 1956. 47 pages. Carnegie Museum, 4400 Forbes Avenue, Pittsburgh 15213. Birding calendar and suggestions for birding trips. Bar graphs with natural areas key, habitat key, and nesting indication. Seven maps, bibliography, index. (JAT)

Birds of Central Pennsylvania. Merrill Wood. 1976. 54 pages. State College Bird Club, 626 W. Nittany Avenue, State College 16801. Annotated list including indication of abundance, status by thirds of each month. Maps identify 21 birding areas. (JAT)

Where to Find Birds in Western Pennsylvania. David Freeland, ed. 1975. 62 pages. Audubon Society of Western Pennsylvania, Beechwood Farms Nature Reserve, 614 Dorseyville Road, Pittsburgh 15238.

Feathers in the Wind. James J. Brett. Hawk Mountain Sanctuary, 1973. 72 pages. Hawk Mountain Sanctuary Association, Route 2, Kempton 19529. Complete information, directions, and description of Hawk Mountain. (JAT)

RHODE ISLAND

Check-list of Rhode Island Birds 1900–1973. Compiled by the Rhode Island Ornithological Club. 1974. 4 pages. Audubon Society of Rhode Island, 40 Bowen Street, Providence 02903.

Field Checklist of Rhode Island Birds. Robert A. Conway. Bulletin No. 1, 1979. 42 pages. Rhode Island Ornithological Club, Audubon Society of Rhode Island, 40 Bowen St., Providence 02903.

SOUTH CAROLINA

South Carolina Bird Life. Alexander Sprunt, Jr., and E. Burnham Chamberlain. 1949; revised 1970. 655 pages. University of South Carolina Press, Columbia 29208. A survey of the natural history of the birds of the state, as complete as available records allowed. Paintings by Francis Lee Jaques, Roger T. Peterson, John Henry Dick, and Edward von S. Dingle. There is also a generous sprinkling of good-to-excellent photographs. (SRD)

Checklist of South Carolina Birds. Compiled by E. Burnham Chamberlain. 1978. 4 pages. Charleston Natural History Society, Charleston Museum, 360 Meeting St., Charleston 29403.

Birds of the Francis Marion National Forest. 1969. 32 pages. Forest Supervisor, U.S. Forest Service, 1801 Main Street, Columbia 29201. Map available separately.

Birds of the Cape Romain National Wildlife Refuge. 1967. 4 pages. Manager, Cape Romain National Wildlife Refuge, Box 191, Route 1, Awendaw 29429.

Birds of the Carolina Sandhills National Wildlife Refuge. 1966. 3 pages. Manager, Carolina Sandhills National Wildlife Refuge, Box 130, Route 2, McBee 29101.

Birding Areas Around Columbia. 1 page. Columbia Audubon Society, P.O. Box 5923, Columbia 29250. Cardboard sheet map showing the location of six good birding sites near Columbia.

SOUTH DAKOTA

Birds of South Dakota, revised edition. William H. Over and Craig S. Thomas. 1946. 142 pages. University of South Dakota Museum, Vermillion 57069. A rather burdensome tome containing, however, valuable information for the student of South Dakota bird life. (SRD)

Birds of the Black Hills. Olin S. Pettingill and Nathaniel R. Whitney. Special Publication No. 1, 1965. 139 pages. Cornell Laboratory of Ornithology, 159 Sapsucker Woods Road, Ithaca, N.Y. 14850. Well-done discussion of the species treated; contains better-than-average information on migration, abundance, and local breeders. Approximately three-fourths of the book concerns breeding species in South Dakota. The remainder deals with birds in the Wyoming section of the Black Hills. (SRD)

Birds of South Dakota, Field Checklist. 1980. 8 pages. South Dakota Ornithologists Union, Route 4, Box 252, Brookings 57006.

Black Hills National Forest Checklist of Birds. 1972. 8 pages. Forest Supervisor, Forest Service Office Building, Custer 57730.

TENNESSEE

Bird-Finding in Tennessee. Michael Lee Bierly. 1980. 225 pages. Published by and available from the author, 3825 Bedford Ave., Nashville 37215. Detailed descriptions of 112 good birding locations with an annotated list of the 342 species recorded in Tennessee. The book also includes a directory of birders from throughout the state, complete with addresses and phone numbers.

Notes on the Birds of Great Smoky Mountains National Park. Arthur Stupka. 1963. 242 pages. University of Tennessee Press, 293 Communications Building, Knoxville 37916. Introduction gives description of area and previous ornithological history. Extensive species accounts. Gazetteer, references, index. (JAT)

Checklist of Birds of Knox County, Tennessee. Alsop. 4 pages. Available free with self-addressed, stamped envelope from Chester Massey and Company, 1301 Hannah Avenue, N.W., Knoxville 37917.

Birds of the Nashville Area. Henry E. Parmer. 1975. 29 pages. Available from ABA Sales. Description of 26 local birding areas near Nashville with lists of species known to occur in each locale. One map. (JAT)

TEXAS

Bird Life of Texas. Harry C. Oberholser and Edgar B. Kincaid, Jr., et al., eds. 1974. 2 vols. 1,057 pages. University of Texas Press, P.O. Box 7819, University Station, Austin 78712. A monumental work incorporating 36 color and 36 black-and-white reproductions and 4 pen-and-ink drawings by Louis Agassiz Fuertes; 2 full-page maps and 480 distributional maps; and 38 habitat photographs. The text includes detailed plumage descriptions, extensive delineation of species' range and seasonal distribution. One of the most valuable and rather remarkable sections deals with the species distribution changes over the years. (SRD)

Birds of Big Bend National Park and Vicinity. Roland H. Wauer. 1973. 223 pages. University of Texas Press, P.O. Box 7819, University Station, Austin 78712. Wauer meant this small volume to be primarily a bird-finding guide. However, it is additionally a guide to the vegetation of Big Bend. Includes illuminating sections on the history of ornithological study in the park area, nesting dates and activities, and migration arrival and departure notes. When the book went to press there were 385 species recorded in the park, which holds the record for parks in the United States. (SRD)

Checklist of the Birds of Texas. L. R. Wolfe. 1975. 128 pages. Texas Ornithological Society, P.O. Box 19581, Houston 77024. Also available from ABA Sales. Detailed species accounts, map, bibliography.

Birds of the Rio Grande Delta Region. L. Irby Davis. 1966; revised 1974. 56 pages. Available from the author, 2502 Keating, Austin 78703. Species accounts and map of area, which includes Mexican part of delta. Species names do not follow the American Ornithologists' Union checklist.

Bird Finding and Naturalist's Guide for the Austin, Texas, Area. E. Kutac and S. C. Caran. 1976. 158 pages. Available from ABA Sales. A bird-finding guide including an annotated checklist, bar graphs, birding areas, and habitat information. (JAT)

Birders' Guide to the Texas Coast. James Lane and John Tveten. 1980. 117 pages. L and P Press, Box 21604, Denver, Colo. 80221. Also available from ABA Sales.

Birders' Guide to the Rio Grande Valley of Texas. James Lane. 1978. 111 pages. L and P Press, Box 21604, Denver, Colo. 80221. Also available from ABA Sales.

Texas Birding Marathon. James A. Tucker. 1971. 27 pages. Available from ABA Sales. Late April birding in Texas.

UTAH

Utah Birds. William H. Behle and Michael L. Perry. Utah Museum of Natural History, 1975. 148 pages. University of Utah Press, Salt Lake City 84112. An up-to-date checklist, including dates, charts of seasonal occurrence, also charts of habitat occurrence. A practical and well-done booklet complete with guide for finding birds in the state. (SRD)

Field Checklist of the Birds of Utah. William H. Behle and Michael L. Perry. 1975. 4 pages. Utah Museum of Natural History, University of Utah, Salt Lake City 84112.

Birds of Bear River Migratory Bird Refuge. 1973. 6 pages. Manager, Bear River National Wildlife Refuge, Box 459, Brigham City 84302.

Seasonal Abundance of Birds on the Bear River Refuge. 1972. 8 pages. Manager, Bear River National Wildlife Refuge, Box 459, Brigham City 84302. Graphs showing monthly relative numbers and maximum numbers for 81 species. (JAT)

Checklist of Birds, Bryce Canyon National Park and Vicinity. R. H. Gerstenberg and R. W. Russell. 1972. 4 pages. Bryce Canyon Natural History Association, Bryce Canyon 84717.

Birds of Zion National Park and Vicinity. R. H. Wauer and Dennis L. Carter. 1965. 92 pages. Zion Natural History Association, Zion National Park, Springdale 84767. Designed primarily as a finding aid but excels in its succinct discussions of habitat and migration and contains an annotated list of more than 230 species reported in the park, plus one map and 22 color plates. (SRD)

VERMONT

Birds of Vermont. Robert N. Spear, Jr., compiler. Green Mountain Audubon Society in cooperation with the State of Vermont Fish and Game Department, 1976. Green Mountain Audubon Society, P.O. Box 33, Burlington 05401. A modest but very nicely executed, updated revision of the 1971 booklet with extensive changes and additions, based on accumulated Vermont bird records. Short species descriptions and status accounts, excellent abundance bar graphs including nesting status and seasonal occurrence information and a comprehensive 4-page chart summary of accidentals and stragglers. Some detail given to birding areas around the state. (SRD)

Checklist for Birds of Vermont. Robert N. Spear. 1965. 40 pages. Cosponsored by the Green Mountain Audubon Society and Vermont Fish and Game Dept. Vermont Fish and Game Department, Montpelier 05602.

Birds of East Central Vermont. Richard B. Farrar. 1973. 106 pages. Vermont Institute of Natural Science, Woodstock 05091. Annotated checklist, plus guide to good birding areas; map and bar graphs. (JAT)

Vermont Daily Field Card. Compiled by Whitney Nichols. 1979. 2 pages. Vermont Institute of Natural Science, Woodstock 05091.

VIRGINIA

Birds of Virginia. Virginia Society of Ornithology. 4 pages. Treasurer, Virginia Society of Ornithology, Department of Biology, College of William and Mary, Williamsburg 23185.

Virginia's Birdlife: An Annotated Checklist. Mrs. Yu Lee Larner, ed. Virginia Society of Ornithology, 1979. 117 pages. Mrs. John Dalmas, Treasurer, Virginia Society of Ornithology, 520 Rainbow Forest Drive, Lynchburg 24502. Distribution and status of 400 species of Virginia birds.

Birds of the Chincoteague National Wildlife Refuge. 1966. 4 pages. Manager, Chincoteague National Wildlife Refuge, Box 62, Chincoteague 23336.

Birds of Back Bay National Wildlife Refuge. 1970. 4 pages. Manager, Back Bay National Wildlife Refuge, Box 6128, Virginia Beach 23456.

Birds of Lynchburg, Virginia, and Vicinity, revised. Ruskin S. Freer. 1973. 104 pages. Lynchburg College Bookstore, Lynchburg College, Lynchburg 24504. Description of area, history of birding in area, annotated bird list, full-page photos. (JAT)

WASHINGTON

Birds of Southeastern Washington. John W. Weber and Earl J. Lanison. 1979. 66 pages. University Press of Idaho, P.O. Box 3368, University Station, Moscow 83843. This distribution checklist includes discussion of recent changes in bird populations as well as notes on geography and climatology of southeastern Washington and neighboring Idaho.

Checklist of the Birds of Washington State. Philip W. Mattocks, Jr., Eugene S. Hunn, and Terence R. Wahl. 1977. 24 pages. Western Field Ornithologists, Inc., 376 Greenwood Beach Road, Tiburon, Calif. 94920. This is a reprint from *Western Birds,* Vol. 7, No. 1, 1976, updating the 1953 Jewett et al. listing through 1974. Each of the 377 species logged is supported by incontestable documentation, and the more than 50 species added since 1953 are annotated. (SRD)

Washington Birds: Their Location and Identification. Earl J. Lanison and Klaus G. Sonnenberg. 1968. 258 pages. Seattle Audubon Society, 714 Joshua Green Building, 1425 4th Avenue, Seattle 98101. Primarily an identification guide with some valuable comments on behavior, but excellent on updating the distributional summaries of the species through 1968. It indicates considerable research on the then-current status of Washington's avifauna. (SRD)

Guide to Bird-finding in Washington. Terence R. Wahl and Dennis Paulson. 1974. 102 pages. Available from ABA Sales. Introduction and description of climate and habitats. Checklist gives status in 5 regions of state, relative abundance for 14 habitats, and indication of breeding. Detailed directions for 93 birding areas. Guide to identification of loons, jaegers, and gulls. (JAT)

Checklist of Washington Birds. Terence R. Wahl. 1972. 6 pages. Whatcom Museum, 121 Prospect Street, Bellingham 98225.

Field Guide to Bird Watching in Skagit County. Tracy Tivel and Jack Adkins. 1976. Map folder. Department of Game, 600 N. Capitol Way, Olympia 98504.

WEST VIRGINIA

List of West Virginia Birds. George A. Hall. 1971. 18 pages. Handlan Chapter, Brooks Bird Club, Inc., Sunrise Museum, 746 Myrtle Rd., Charlestown, 25314.

List of Winter Birds of Oglebay Park. 1969. 2 pages. Nature Education Department, Oglebay Institute, Wheeling 26003.

WISCONSIN

Birds of Wisconsin. Owen J. Gromme. 1963. 219 pages. University of Wisconsin Press, P.O. Box 1379, Madison 53701. Useful as the distribution is statewide. Contains bar graphs and rather telegraphic discussion of species abundance. (SRD)

Field Check-list of Birds of Wisconsin. 1978. 4 pages. Wisconsin Society for Ornithology, Inc., Supply Department, 246 N. High St., Randolph 53956.

Wisconsin Birds: A Checklist with Migration Charts. N. R. Barger, Jr., et al. 1960. 32 pages. Wisconsin Society for Ornithology, Inc., Supply Department, 246 N. High St., Randolph 53956. Gives status, relative abundance for 4 portions of state; also bar graphs. Space for 10 days' records. Map with list of 24 birding areas, bibliography. (JAT)

Wisconsin's Favorite Bird Haunts. Daryl Tessen, ed. 1976. 334 pages. Wisconsin Society for Ornithology, Inc., Supply Department, 246 N. High St., Randolph 53956. Directions to 90 birding areas, with numerous maps.

Wisconsin Favorite Bird Haunts—Supplement. Daryl Tessen, ed. 1979. 424 pages. Wisconsin Society for Ornithology, Inc., Supply Department, 246 N. High St., Randolph 53956. Includes twelve additional birding sites and updated information about many of the birding localities described in the 1976 title.

Birds of Prey of Wisconsin. Francis Hamerstrom. 1972. 64 pages. Department of Natural Resources, Madison 54301.

Birds of the Whitnall Park Area (Milwaukee). 1972. 15 pages. Available free with self-addressed, stamped envelope from Boerner Botanical Gardens, Whitnall Park, 5879 S. 92nd Street, Hales Corner 53130. The data for each species are their status and relative abundance locally and in state, a bar graph, and the area of their usual local occurrence. Bibliography. (JAT)

WYOMING

Checklist: Birds of Wyoming. Oliver T. Scott. 1975. 4 pages. Available free with self-addressed, stamped envelope from the author, Box 5360, Alcove Route, Casper 82601.

Canada

ALBERTA

Birds of Alberta, 2nd edition. Walter Raymond Salt. 1976. 498 pages. Hurtig Publishers, 10560 105th Street, Edmonton T6E 4T5. Excellent reference that has suffered little with the passage of time. Should be invaluable to students of the bird life of Alberta. (SRD)

Check-list of Alberta Birds. H. C. Smith. 1975. 6 pages. Provincial Museum of Alberta, 12845-102nd Ave. Edmonton T5N OM6. A field list of the 328 species of birds that occur in Alberta. Revised list in preparation.

Checklist of the Birds of the Calgary Region. Stuart Alexander et al. 1969. 4 pages. Calgary Field Naturalists' Society, P.O. Box 981, Calgary T2P 2K4.

Parkways of the Canadian Rockies. Brian Patton. 1975. 192 pages. Summerthrough Ltd., Box 1420, Banff T0L OCO.

BRITISH COLUMBIA

Where to Find Birds in British Columbia. David M. Mark. 1978. 72 pages. Krestrel Press, P.O. Box 2054, New Westminster V3L 5A3. Directions to more than 40 birding areas. Eleven maps. Includes a checklist of the birds of British Columbia. (JAT)

Bird Report for Southern Vancouver Island. J. B. Tatus, ed. 1971. 72 pages. Available from the author, 305, 1680 Poplar Street, Victoria. Issued annually, this complete record of the previous year gives migration dates, census results, breeding records, and accounts of rarities. Section on "The Ornithological Year." Complete accounts of each species occurrence throughout the year. Maps, illustrations, photos. (JAT)

Checklist of British Columbia Birds. 1978. Museum Gift Shop, British Columbia Provincial Museum, Victoria V8V 1X4.

Birder's Guide to Victoria, BC. 1973. 79 pages. Museum Gift Shop, British Columbia Provincial Museum, Victoria V8V 1X4.

Birds of Vancouver Island. David Stirling. 1972. 26 pages. Available from the author, 330–1870 McKenzie Ave., Victoria V8N 4X3. Discussion of topography, climate, and habitat changes. Species accounts. (JAT)

Vancouver Checklist. John Toochin et. al. 1 page. Vancouver Natural History Society, P.O. Box 3021, Vancouver V6B 3X5.

MANITOBA

Field Checklist of Manitoba Birds. 1974. 6 pages. Manitoba Museum of Man and Nature, 147 Jones Avenue, Winnipeg 2.

Birds of the Churchill Region, Manitoba. Joseph Jehl and Blanche Smith. Special Publication No. 1, 1970. 88 pages. Manitoba Museum of Man and Nature, 147 Jones Avenue, Winnipeg 2. Description of area, ornithological history, recent changes in environment, and section on birding at Churchill. Very detailed species accounts include egg dates when known, and arrival and departure dates. Seventeen photos and drawings, 3 maps, and joint index and checklist. Extensive bibliography. (JAT)

Birds of Alberta, Saskatchewan, and Manitoba. David H. Hancock and Jim Woodford. 1973. General Publishing Co., Don Mills, Ontario M3B 2W7.

NEW BRUNSWICK

Birds of New Brunswick. William Austin Squires. 1976. 220 pages. New Brunswick Museum, 277 Douglas Avenue, St. John's E2K 1E5. Contains much information on geography and climate, bird banding data, sources of unpublished information, migration routes, and an excellent annotated list. The bibliography is above average. Recommended. (SRD)

Finding Birds Around Saint John, David S. Christie. 1978. 28 pages. Sales Department, New Brunswick Museum, 277 Douglas Avenue, Saint John's, E2K 1E5.

Field Check-list, New Brunswick Birds. 1975. 4 pages. Sales Department, New Brunswick Museum, 277 Douglas Avenue, St. John's E2K 1E5.

Check-list of Birds: Nova Scotia—New Brunswick Border Region. 1974. 4 pages. Canadian Wildlife Service, Department of the Environment, Box 1590, Sackville E0A 3C0.

Check-list of Birds: Cape Jourimain National Wildlife Area [Northumberland Strait]. 6 pages. Canadian Wildlife Service, Department of the Environment, Box 1590, Sackville E0A 3C0.

Check-list of Birds: Tintamarre National Wildlife Area [Sackville vicinity]. 6 pages. Canadian Wildlife Service, Department of the Environment, Box 1590, Sackville E0A 3C0.

Birds of St. Andrews. Willa MacCoubrey and Tom Moffat. 1976. 6 pages. Sunbury Shores Arts and Nature Center, 139 Water Street, Box 100, St. Andrews E0G 2X0.

NEWFOUNDLAND

Birds of Newfoundland. H. S. Peters and T. D. Burleigh. 1951. 431 pages. Department of Natural Resources, St. John's A1A 1P9. The only comprehensive publication on Newfoundland birds. (SRD)

Checklist of Newfoundland Birds. Leslie M. Tuck. 1967. 8 pages. Canadian Wildlife Service, P.O. Box 6028, Building 304, St. John's A1C 5X8.

Checklist of Birds: Terra-Nova National Park. 6 pages. Park Naturalist, Terra-Nova National Park, National and Historical Parks Branch, Glovertown.

NORTHWEST TERRITORIES

Birds of South East Victoria Island and Adjacent Small Islands. David F. Parmelee et al. National Museum of Canada Bulletin 222, Biological Series No. 78, 1967. 229 pages. National Museums of Canada, Marketing Services Division, 360 Lisgar Street, Ottawa, Ontario K1A 0M8. Introduction, historical notes, descriptions of climate and nesting habitats, and extensive species accounts. List of mammals and bibliography. Many photos. Available on microfiche from Micromedia Limited, 144 Front Street, West Toronto, Ontario M5T 2L7. (JAT)

NOVA SCOTIA

Birds of Nova Scotia. Robie W. Tufts. 1961. 481 pages. Nova Scotia Museum, 1747 Summer Street, Halifax B3H 3A6. An updated version was published in 1974. Contains excellent plates and is recommended for a survey of the birds of Nova Scotia. It is the only major work available on that area and is a worthwhile contribution to the literature. (SRD)

Nova Scotia Bird Society Check List. 1970. 4 pages. Nova Scotia Bird Society, c/o Nova Scotia Museum, 1747 Summer Street, Halifax B3H 3A6.

Where to Find Birds in Nova Scotia. P. R. Dobson and C. R. K. Allen. 1976. 37 pages. Nova Scotia Bird Society, c/o Nova Scotia Museum, 1747 Summer Street, Halifax B3H 3A6. Introduction, birding calendar, and information on weather. Fifteen pages of directions to birding areas, with expected species. Index of most-sought-after species. Contains list of county representatives of Nova Scotia Bird Society, with addresses and phone numbers. Designed to be used in conjunction with official highway map and Nova Scotia Tour Book, both available from any tourist bureau. (Latter contains an annotated list of Nova Scotia birds.) (JAT)

Check List: Birds of Cape Breton Highlands National Park. 1974. 16 pages. Park Naturalist, Cape Breton Highlands National Park, Ingonish Beach, B0C 1L0. English and French.

ONTARIO

Annotated Checklist of the Birds of Ontario. R. D. James, P. L. McLaren, and J. C. Barlow. 1976. 75 pages. Royal Ontario Museum, Life Sciences Miscellaneous Publications, Toronto. Also available from ABA Sales. A tight compilation of the 394 living, 1 extinct, and 34 "hypothetical" bird species recorded in Ontario. Included are 267 current or former breeders and 12 summer residents (judged as probable but undocumented breeders). Species accounts include breeding and migratory status, seasonal occurrence and frequency, relative abundance, distribution, egg dates, and subspecies distribution. (SRD)

Field Checklist of Birds, 1980. 6 pages. Federation of Ontario Naturalists, 355 Lesmill Road, Don Mills M3B 2WB.

Bird Finding Guide to the Toronto Region. Clive Goodwin. 1979. 97 pages. Toronto Field Naturalists, 83 Joicey Boulevard, Toronto M5M 2T4. Directions to 134 birding locations within a 30-mile radius of Toronto. Seasonal abundance lists for 305 species. A good reference for southern Ontario.

Bird Migration Chart: Toronto Field Naturalists' Club. R. M. Sanders and J. L. Baillie.

1968. 32 pages. Secretary, Toronto Field Naturalists, 83 Joicey Boulevard, Toronto M5M 2T4. Pocket-sized companion to *Toronto Birdfinding Guide*. Bird graphs do not give relative abundance, but the number of years in which each species was seen by each author is given. (JAT)

Birds of Prince Edward County. Terry Sprague. 1969. 215 pages. Prince Edward Region Conservation Authority, P.O. Box 3032, Picton K0K 2T0. Description of area, species accounts, map, and bibliography.

Birds of the Oshawa–Lake Scugog Region. Ronald G. Tozer and James M. Richards. 1974. 384 pages. James M. Richards, R.R. #2, Orono L0B 1M0. Description of area, detailed species accounts include useful information about breeding biology, time and locations for observing birds in the region. Also map, extensive bibliography, several appendixes.

Birds of the Hudson Bay Lowlands of Ontario, revised. J. P. Prevett. 1979. Ontario, revised. Ministry of Natural Resources. District Manager, Box 190, Moosonee, Ontario P0C 140.

PRINCE EDWARD ISLAND

Provisional Checklist of Birds, Prince Edward Island National Park. 1964. 8 pages. Prince Edward Island National Park, P.O. Box 487, Charlottetown C1A 7N8.

Prince Edward Island Field Checklist of Birds. 1978. 8 pages. Department of Tourism, Parks and Conservation, P.O. Box 2000, Charlottetown C1A 7N8.

QUEBEC

Field Check-list of Birds in the Montreal Area. 1978. 8 pages. Province of Quebec Society for the Protection of Birds, Inc., P.O. Box 43, Station B, Montreal H3B 3J5. English and French list of 267 species known to have nested or occurred five times or more in the Montreal area.

In Search of Birds. Robert Carswell. 1979. 10 pages. Province of Quebec Society for the Protection of Birds, P.O. Box 43, Station B, Montreal H3B 3J5.

Status and Distribution of Birds in Southern Québec. Normand David. 1980. 213 pages. Chahiers d'ornithologie Victor-Gaboriault, no. 4. Charlesbourg, Québec. Available from Club des ornithologues due Québec, a/s Fédération québecoise du loisir scientifique, 1415 est, rue Jarry, Montréal, Québec H2E 2Z7. Distribution and status of 408 bird species from southern Quebec (south of latitude 52 00°N). The book also includes a useful bibliography.

Bird Check-list of the Percé Area. 1974. 8 pages. Perce Wildlife Centre, CP/Box 190, Perce.

Sea-birds of Bonaventure Island. 1973. 20 pages. Government of Canada Publication No. CW66-4173. Publishing Centre, Printing and Publishing, Supply and Services Canada, Ottawa, Ontario KIA 059. Illustrated booklet describing the island and its magnificent seabird colonies.

Birds of the Ungava Peninsula. Francis Harper. Miscellaneous Publication No. 17 of the University of Kansas Museum of Natural History, 1958. 171 pages. Arctic Institute of North America, 1619 New Hampshire Avenue, N.W., Washington, D.C. 20009. An account of 1953 investigations. Covers physiography, climate, vegetation, and factors influencing bird distribution. Extensive species accounts, many photos and drawings. Bibliography, index. (JAT)

SASKATCHEWAN

Field Checklist of Saskatchewan Birds. Saskatchewan Museum of Natural History. 1969. 4 pages. The Blue Jay Bookshop, Box 1121, Regina S4P 3B4.

Birds of the Gainsborough–Lyleton Region [Saskatchewan and Manitoba]. Richard W. Knapton. Saskatchewan Natural History Society Special Publication No. 10, 1979. 72 pages. The Blue Jay Bookshop, Box 1121, Regina S4P 3B4. An annotated species list for extreme southeastern Saskatchewan and neighboring Manitoba, with dates of occurrence for 246 species known to this prairie region.

Birds of Moose Mountain Saskatchewan. Robert W. Nero and M. Ross Lein. Saskatchewan Natural History Society Special Publication No. 7, 1971. 55 pages. The Blue Jay Bookshop, Box 1121, Regina S4P 3B4. Annotated list discussing abundance and dates of occurrence of birds within Moose Mountain Provincial Park and vicinity in extreme southeastern Saskatchewan.

Birds of Northeastern Saskatchewan. Robert W. Nero. Saskatchewan Natural History Society Special Publication No. 6, 1967. 96 pages. The Blue Jay Bookshop, Box 1121, Regina S4P 3B4. Introduction and description of study areas. Discussion of avifauna, with previous records and species accounts. Photos, references. (JAT)

Birds of the Qu'Appelle, 1857–1979. E. Manley Callin. Saskatchewan Natural History Society Special Publication No. 13, 1980. 168 pages. The Blue Jay Bookshop, Box 1121, Regina S4P 3B4. Species accounts for 287 species inhabiting the Qu'Appelle river valley in southcentral Saskatchewan, providing details on abundance and seasonal occurrence.

Birds of Regina. Margaret Belcher. Saskatchewan Natural History Society, Special Publication No. 12, 1980. 151 pages. The Blue Jay Bookshop, Box 1121, Regina S4P 3B4. Species accounts for 295 species known to occur in this southcentral region of Saskatchewan. Excellent black-and-white photos of some of the characteristic plains habitat and its most notable residents.

Birds of the Rosetown–Biggar District. Wayne E. Renaud and Don H. Renaud. Saskatchewan Natural History Society Special Publication No. 9, 1975. The Blue Jay Bookshop, Box 1121, Regina S4P 3B4. A well-documented account of the bird fauna in Saskatchewan between the North and South rivers, with major emphasis on the breeding status and migratory movements of 111 established breeders of the 236 species definitely recorded in the area. A testimony to the contribution amateurs can make to ornithology. (SRD)

YUKON

List of Yukon Birds and Those of the Canol Road. A. L. Rand. National Museums of Canada Bulletin No. 105, Biological Series No. 33, 1946. 76 pages. National Museums of Canada, Ottawa, Ontario. Available on microfiche from Micromedia Limited, 144 Front St. W., Toronto, Ontario M5J 2L7. Dated, but contains copious records and field accounts. The synopsis of former avifauna and previous work done in the Yukon is noteworthy, especially historically. (SRD)

Field Check-list of the Birds of the Yukon Territory. Game Branch, Yukon Territorial Government and the Yukon Conservation Society. 6 pages. Game Branch, Yukon Territorial Government, Box 2703, Whitehorse. Foldout field card giving abundance and north–south distribution in Yukon Territory.

Distribution and Abundance of Birds on the Arctic Coastal Plain of Northern Yukon and Adjacent Northwest Territories, 1971–1976. Richard E. Salter et al. 1980. 20 pages. Reprinted from *Canadian Field-Naturalist,* Vol. 94, No. 3, July–September 1980. Available from LGL Limited, Environmental Research Associates, Suite 201, 10110, 124 Street, Edmonton T5N 1P6. Map, two habitat photos, and annotated list of 122 species includes abundance, dates of occurrence and breeding, along with other notes of interest.

Bermuda

Checklist and Guide to the Birds of Bermuda. David B. Wingate. 1973. 36 pages. Available from the author, Department of Agriculture and Fisheries, Botanical Gardens, P.O. Box 834, Hamilton 5. Tips on bird-finding and species accounts for the 20 resident bird species. The checklist consists primarily of a calendar chart showing seasonal distribution and abundance for the 320 species known from Bermuda.

Mexico

Field Guide to Mexican Birds. Roger Tory Peterson and Edward L. Chalif. 1973. 298 pages. Houghton Mifflin Co., 2 Park Street, Boston, Mass. 02107. Description of field marks, similar species, range, habitat, and sometimes voice for all species in Mexico, Guatemala, Belize, and El Salvador. Forty-eight excellent color plates with arrows pointing to diagnostic features.

Birds of Mexico. Emmet Reid Blake. 1953. 644 pages. University of Chicago Press, 5801 Ellis Avenue, Chicago, Ill. 60637. The first field guide for Mexico, it contains species accounts and useful keys as well as line drawings of selected species. No color plates.

Field Guide to the Birds of Mexico. Ernest P. Edwards. 1972. 300 pages. J. P. Bell Co., Inc., Lynchburg, Va. Also available from the author, Sweet Briar, Va. 24595. Species accounts including abundance, habitat, and identification descriptions in English and Spanish. Color plates of selected species. Covers Mexico and countries south through Nicaragua.

Bird Finding in Mexico, 2nd edition. Ernest P. Edwards. 1968. 282 pages. 1976 Supplement, 135 pages. J. P. Bell Co., Inc., Lynchburg, Va. Also available from the author, Sweet Briar, Va. 24595. Bird-finding locations throughout Mexico giving travel directions, accommodations, and species to look for.

West Indies

Birds of the West Indies, 3rd edition. James Bond. 1971. 256 pages. Houghton Mifflin Co., 2 Park Street, Boston, Mass. 02107. Including Bahamas, Cuba, Jamaica, Dominican Republic, Haiti, Puerto Rico, and Lesser Antilles. In field-guide format, this guide provides local names, descriptions, habitat, range, and some discussion of nests, eggs, and voice. Line drawings and color illustrations document most resident species.

Birds of New Providence and the Bahama Islands. P. G. C. Brudenell-Bruce. 1975. 142 pages. Taplinger Publishing Co., 200 Park Avenue S., New York, N.Y. 10003. Brief discussion of effects of human occupation, climate, and vegetation on the distribution of Bahamian birds. Species accounts include status, description, voice, and nesting habits. Black-and-white and color plates illustrate resident species.

Guide to the Birds of Trinidad and Tobago. Richard ffrench. 1976. 470 pages. Harrowood Books, Valley Forge, Pa. 19481. Field-guide format with useful discussion of the Trinidad and Tobago environment showing influences on behavior and bird distribution. Notes on breeding distribution and migration with thorough species accounts. Information about habitat, status, range, description, measurements, voice, food, nesting, and details about behavior. A model for tropical field guides. Color plates.

Catalogo de las Aves de Cuba. Orlando H. Garrido and Florentino Garcia Montana. 1975. 149 pages. Academia de Ciencias de Cuba, Havana, Cuba. Latin, English, and Spanish names organized by family. Spanish text. No illustrations.

Annotated Checklist of the Birds, Mammals and Amphibians of the Virgin Islands and Puerto Rico. Richard Philibosian and John A. Yntema. 1977. 48 pages. Information Services, P.O. Box 305, Frederiksted, St. Croix, U.S. Virgin Islands 00840. Three hundred ninety-eight living and extinct species. Charts indicate breeding or visitor status and seasons of occurrence.

Aves de la Republica Dominicana. Annabelle Stockton de Dod. 1978. 332 pages. Museo National de Historia Natural, Santo Domingo, Republica Dominicana. Color and line drawings illustrate this first treatment of Dominican Republic birds. Approximately 2 pages of Spanish text detail behavior and field characteristics about each of 136 species. Range maps and a checklist of the 210 known species make this a valuable guide for anyone interested in the birds of this tropical island.

BIRDS OF THE WORLD

The eight hundred-plus bird species that are found in North America are a good sample of the more than eighty-six hundred species that occur on Earth. However, North America's birds represent only 75 of the world's 176 living bird families. With this in mind, it is useful to have an overview of the remarkable variety of birds that occur in distant parts of the globe, some of which are especially bizarre in comparison to the relatively somber-plumaged birds of North America.

Recommended

Bird Families of the World. D. J. O. Harrison, Editor. New York: Harry N. Abrams, 1978. 264 pages, hardcover.

Only the inspired art of an illustrator such as Ad Cameron could begin to portray the variety of shapes, colors, and behaviors expressed by the world's birds. Combine this talent with the expertise of forty-one bird authorities, and the result is *Bird Families of the World.* Published in large format, the readable text is packed with useful information. The book describes the many fossil families and all living bird families by first presenting a brief overview of characteristics and the variety within the family. Each family description contains subheadings that describe distribution, feeding habits, and nesting behavior. For some families there is also a statement describing the economic value of the group to humanity.

Families of Birds. Oliver L. Austin, Jr. New York: Golden Press, 1971. 200 pages, hardcover and paperback.

Within its pages, this compact book summarizes 172 living and 36 fossil families of birds. This concise treatment is an excellent framework for learning bird classification. Although its small format does not permit either the depth of information or abundance of illustrations found in *Bird Families of the World,* the Austin guide does give the reader a feeling for the variety of birds found throughout the world. This is accomplished largely by Arthur Singer's colorful bird portraits.

With the exception of fossil families, each family account gives number of species, size of range, world distribution, sexual differences, and nesting habits. Introductory pages contain useful illustrations of the faunal regions of the world, a geologic table illustrating the appearance and abundance of major bird orders, and a family tree showing the relationships of living birds to their fossil ancestors.

The World Atlas of Birds. Peter Scott, Editor. New York: Random House, 1974. 272 pages, hardcover.

This abundantly illustrated book brings an ecological approach to a world survey of birds. Authorities from many countries describe their home region and the birds that occupy typical ecological communities. The emphasis of the book is a habitat review of each continent and discussion of how geology, climate, and history have shaped the birds of each region. *The World Atlas of Birds* is a refreshing approach to the traditional family classification of birds, but it is not a substitute for learning the relationships between bird families.

In addition to the habitat review, there is also a brief introductory section on bird biology with discussion of such topics as bird evolution, adaptation, flight, feeding habits, and migration. The book concludes with a brief world survey of all bird orders and families.

Other Choices

Birds of the World. Oliver L. Austin, Jr. New York: Golden Press, 1961. 316 pages, hardcover (out of print).

With large, colorful illustrations by Arthur Singer, this is a dynamic survey of 27 orders and 155 bird families. Much of the text is written in first-person narrative as Austin describes his experiences with representative species. This provides a highly readable, conversational quality to the text.

The Dell Encyclopedia of Birds. Bertel Bruun. New York: Dell Publishing, 1974. 240 pages, paperback.

This inexpensive handbook not only introduces 29 orders and 172 families of world birds, but also contains definitions and brief descriptions of many aspects of bird behavior and anatomy. Abundantly illustrated by Paul Singer's colorful art, this is a convenient reference to many frequently used ornithology terms.

BIRD SONG RECORDINGS

There are two basic types of bird song recordings—those that function primarily as aids for learning to identify bird sounds, and those that convey the pleasure of just listening to bird music. Useful records for identifying bird sounds should contain samples of songs and calls extensive enough to illustrate the varied vocabulary and dialects that birds exhibit. The recommended choices provide good samples of bird song and include the species most frequently encountered by beginning bird-watchers.

Recommended

Common Bird Songs (1967), *Songs of Eastern Birds* (1970), *Songs of Western Birds* (1971), and *Bird Song and Bird Behavior* (1972). Donald J. Borror. Four 12-inch, 33⅓ rpm, monaural record albums, each accompanied by a manual (27 pages for *Common Bird Songs,* 32 pages for the other three). New York: Dover Publications.

This four-volume series is an excellent aid for learning to recognize the songs and calls of common birds. Each album not only contains an ample selection of field-recorded songs for each species, but also a good sampling of characteristic variations in song and call for each entry. The illustrated manual that accompanies each album nicely complements the record by providing excellent line drawings and interesting text about birds featured on the recordings.

Common Bird Songs is the most useful identification album available for beginning bird-watchers in eastern North America. It includes the songs and calls of sixty common species, such as the common crow, blue jay, white-breasted nuthatch, and wood thrush. Birds with similar songs are grouped together for convenient comparison.

Songs of Eastern Birds presents the vocalizations of sixty additional eastern birds, many of which live in forests and marshlands where they are seldom seen but often heard. *Songs of Western Birds* presents the songs of sixty additional bird species found in the western United States.

Bird Song and Bird Behavior explains the function of songs and reminds us that bird song communicates much more than just the name of the bird. Vocalizations play an important role in understanding bird behavior, and this album provides the necessary interpretation to better understand bird communication. The manual that accompanies this innovative record contains photographs of equipment used to make and analyze bird songs and explains how to read sound spectographs.

A Day in Algonquin Park. William W. H. Gunn. Sounds of Nature, Vol. 2. Published by Houghton Mifflin Co. in the United States and elsewhere and by the Federation of Ontario Naturalists in Canada. Twelve-inch, 33⅓ rpm record.

This thirty-minute album presents many of the voices characteristic of the northern forests, lakes, and marshes in Ontario. Side 1 begins with the dawn voice of the white-throated sparrow and proceeds through a day-long canoeing experience past the varied sounds of bullfrogs, chickadees, and even an occasional mosquito. Side 2 presents evening songs such as those of the whippoorwill and nighthawk. Spring peepers and American toads also add their voices to the dusk chorus. The album ends with a haunting symphony of loon music. The wild voices of Algonquin Park are uninterrupted by human narration throughout the album, except for a brief summary to review and identify the sounds at the conclusion of Side 2.

An Evening in Sapsucker Woods. Peter Paul Kellogg and Arthur A. Allen. Sounds of Nature, 1958. Published by Houghton Mifflin Co. in the United States and elsewhere and by the Federation of Ontario Naturalists in Canada. Ten-inch, 33⅓ rpm record.

Sapsucker Woods is the sanctuary of the Cornell Laboratory of Ornithology in Ithaca, New York. This classic recording by Drs. Allen and Kellogg presents the voices of twenty-seven birds and five types of frog and toad that sing during the evenings within the forests and marshlands of the northeastern United States. On Side 1 Dr. Allen identifies the bird and frog sounds most commonly heard in Sapsucker Woods. Side 2 begins with the liquid song of the veery and continues to the flutelike voice of the wood thrush. Without human interruption, Side 2 continues with choruses of frogs and concludes with the bass voice of a barred owl. The absence of narration gives the listener a chance to review many of the sounds identified from Side 1. *An Evening in Sapsucker Woods* will revive the warmth and life of a spring night long into the quiet of winter.

Other Choices

Warblers, Finches, and *Thrushes, Wrens, and Mockingbirds.* Vols. 4, 6, and 8, respectively, of Sounds of Nature Series. Donald J. Borror and William W. H. Gunn. Each a 12-inch, 33⅓ rpm, monaural record. Published by Houghton Mifflin Co. in the United States and elsewhere and by the Federation of Ontario Naturalists in Canada.

Bird-watchers already familiar with the songs and calls of common birds will find these volumes valuable, because they include nearly all North American species within the featured groups and present an excellent sample of songs from several different locations.

Warblers presents the songs of thirty-eight species that regularly breed in eastern North America. Similar-sounding species are grouped together to permit convenient comparisons. The depth of this series is illustrated by the album entitled *Finches,* which includes forty-three species of finches, sparrows, and grosbeaks. This album contains approximately 400 songs from 226 different individual birds. *Thrushes, Wrens, and Mockingbirds* presents an extensive collection of the songs and calls of all seventeen species that breed in eastern North America.

A Field Guide to Bird Songs. (Eastern and Central North America). Peter Paul Kellogg and Arthur A. Allen, Editors. Boston: Houghton Mifflin Co. Two 12-inch, 33⅓ rpm records or three tape cassettes. Entries correspond by page number to the 2nd revised edition of *A Field Guide to the Birds* by Roger Tory Peterson. Page numbers do not correspond to the new, completely revised 4th edition.

A Field Guide to Western Bird Songs. Peter Paul Kellogg, Editor. Boston: Houghton Mifflin Co., 1962. Three 12-inch, 33⅓ rpm records or four tape cassettes. Entries correspond by page number to the 10th or later printings of the 2nd edition of *A Field Guide to Western Birds* by Roger Tory Peterson.

These companion records to two of Roger Tory Peterson's field guides contain a vocal sample from nearly all North American birds. The eastern album contains recordings from more than three hundred species, while the western album presents songs from more than five hundred species. These impressive numbers make the recordings useful, as one can almost certainly find at least a brief recording of any species that occurs in North America. The records are keyed to the guides by a narrator who announces the name of each bird, followed by the corresponding page number in the field guide.

The principal disadvantage of this series is the short burst of song allotted to each species. Considering that many birds have regional dialects and most can sing a variety of songs and produce many calls, it is often misleading to think the brief selection included is representative of the species.

Beautiful Bird Songs of the World. Ithaca: National Audubon Society and Cornell Laboratory of Ornithology, 1977. Two 12-inch, 33⅓ rpm monaural records and a 12-page booklet with paintings by Arthur Singer.

Fifty species of birds with exceptionally beautiful songs are featured in this two-record album, including the voices of the nightingale, song thrush, and woodlark of Europe. The album also features the cardinal, wood thrush, and mockingbird of North America along with a selection of talented avian voices from Australia, New Zealand, and Asia. The two-record format provides ample time for each bird to perform a good sample of its song repertoire. While listening to the recordings, you can read the enclosed booklet, which contains pleasing illustrations by Arthur Singer and a brief account of the behavior and nesting habits of each species.

Bird Songs in Your Garden (1961), *Songbirds of America* (1954), and *Dawn in a Duckblind* (1963). Three 10-inch 33⅓ rpm records, each accompanied by a booklet (24 pages for *Bird Songs in Your Garden,* 28 pages for the other two). Published by Houghton Mifflin Co. in the United States and elsewhere and by the Federation of Ontario Naturalists in Canada. Available from the Cornell Laboratory of Ornithology.

These albums contain much more than just the voices of birds. This classic series combines color photography with bird song recordings and narration. In addition to the photography and sound, each "bookalbum" contains factual accounts of selected species and an introductory discussion about selected bird biology topics. Although some of the text is dated, these volumes remain good introductions for the beginning bird-watcher who wants to know more about the most common birds of backyards and wetlands.

The book portion of *Songbirds of America* is prefaced by discussion of such topics as bird-human interactions, biology of bird colors, nesting habits, and function of bird song. This bookalbum presents the songs of twenty-four common eastern land birds, such as Carolina wren, robin, red-winged blackbird, mockingbird, and cardinal. Each of the twenty-four birds is illustrated by a color photograph, and there is accompanying information about range, size, habits, and song.

Bird Songs in Your Garden presents twenty-five eastern birds. These are also familiar garden birds, although many of those included here are also featured in *Songbirds of America.* In addition to natural history discussion of each species, photographs, and songs, this bookalbum contains several pages of bird-attracting information, such as suggested plantings for attracting hummingbirds, designs for birdhouses, bird feeders, and birdbaths.

Dawn in a Duckblind features the bird sounds of marshes and lakes. It is an album that will have special appeal to waterfowl hunters and bird-watchers who enjoy the sight of ducks and geese and who associate their sounds with wild wetlands. In addition to fifteen species of waterfowl, the album also presents the voices of other wetland birds, such as pied-billed grebe, common loon, red-winged blackbird, common coot, killdeer, and Virginia rail. The text, however, includes only discussion and photographs of waterfowl, omitting the other marsh birds. A useful feature of the album section is a listing of some of the more important national waterfowl refuges with information on size, location, and periods of waterfowl concentration.

Audible Audubon. National Audubon Society, 1977. Eighty-one 3 × 5-inch microphonograph cards, player, carrycase, and 27 cards. Two sets of 27 additional cards. Available from the National Audubon Society.

The *Audible Audubon* collection is a convenient way to review bird songs at home and in the field. This is the only unit designed for field use and instant replay against the sounds that are actually heard. The system is especially useful for field trip leaders who want to review bird songs and the appearance of the birds as they are encountered in the field. Those interested in attracting birds will find that the player is not usually loud enough to evoke a territorial defense response. Each card features a different species, with a painting by Allan Brooks or Roger Tory Peterson on the front and useful information compiled by Audubon staff on the back. A microrecording with a selection of bird song and narration is permanently attached to the back of each card, and these are inserted one at a time into the hand-sized, battery-operated microphonograph player.

A

SOURCES FOR BINOCULARS AND OTHER BIRD-WATCHING SUPPLIES

BINOCULAR AND SPOTTING SCOPE RETAILERS AND REPAIR SERVICES

Bausch & Lomb Optics Center, 1400 N. Goodman Street, Rochester, N.Y. 14602. Repair service only.

Birding, P.O. Box 5, Amsterdam, N.Y. 12010.

Bushnell Optical Co., 2828 E. Foothill Boulevard, Pasadena, Calif. 91107. Bushnell, Bausch & Lomb binoculars, spotting scopes, tripods.

Carl Zeiss, Inc., 444 Fifth Avenue, New York, N.Y. 10018. Zeiss binoculars.

Celestron International, 2835 Columbia Street, Box 3578AB, Torrance, Calif. 90503. Celestron spotting scopes, telescopes.

Danley's, P.O. Box 600B, Schenectady, N.Y. 12301.

E. Leitz Co., Rockleigh, N.J. 07647. Leitz binoculars.

47th Street Photo, 36 E. 19th Street, New York, N.Y. 10003.

G. L. Hebard Optics, 125129 Public Square, Knoxville, Ill. 61448.

Massachusetts Audubon Society Gift Shop, S. Great Road, Lincoln, Mass. 01773. Catalog fee.

Mirakel Optical Co., Inc., 331 Mansion Street, West Coxsackie, N.Y. 12192. Binoculars, spotting scopes, repair service.

National Audubon Society, The Booknest, Western Education Center, 376 Greenwood Beach Road, Tiburon, Calif. 94920. Binoculars, spotting scopes.

Nature Shop, 16 Holmes Street, Mystic, Conn. 06355.

Nikon, Inc., 623 Stewart Avenue, Garden City, N.Y. 11530. Nikon binoculars.

Questar Corporation, R.D. 1, New Hope, Pa. 18938. Questar telescopes, telephoto lenses.

Swift Instruments, Inc., 952 Dorchester Avenue, Boston, Mass. 02125. Swift binoculars, spotting scopes, tripods.

Tasco Sales, Inc., 1075 N.W. 71st Street, Miami, Fla. 33138. Tasco binoculars, spotting scopes, tripods.

Thos. Manzetta, 61 Hoffman Avenue, Elmont, Long Island, N.Y. 11003.

CAMERA AND TAPE RECORDER RETAILERS

Canon USA, Inc., 10 Nevada Drive, Lake Success, N.Y. 11040. Canon telephoto lenses, cameras.

Craig Corp., 921 W. Artesia Boulevard, Compton, Calif. 90220. Craig tape recorders.

E. Leitz Co., Rockleigh, N.J. 07647.

47th Street Photo, 36 E. 19th Street, New York, N.Y. 10003.

Garden Camera, 345 Seventh Ave., New York, N.Y. 10001.

JVC America Company, 58–75 Queens Midtown Expressway, Maspeth, N.Y. 11378. JVC tape recorders, microphones.

Minolta Corp., 101 Williams Drive, Ramsey, N.J. 07446. Telephoto lenses, cameras.

Nikon, Inc., 623 Stewart Avenue, Garden City, N.Y. 11530. Nikon binoculars, telephoto lenses.

Panasonic Co., 1 Panasonic Way, Secaucus, N.J. 07094. Panasonic tape recorders, microphones.

Sennheiser Electronics Corp., 10 W. 37th Street, New York, N.Y. 10018. Sennheiser shotgun microphones.

Shure Brothers, Inc., 222 Hartrey Avenue, Evanston, Ill. 60204. Shure microphones.

Sony Corporation of America, 9 W. 57th Street, New York, N.Y. 10019. Sony cassette tape recorders, microphones.

Superscope, Inc., 20525 Nordhoff Street, Chatsworth, Calif. 91311. Superscope tape recorders, microphones.

Uher Corp., 612 S. Hindry Avenue, Inglewood, Calif. 90301. Uher tape recorders, microphones.

Vivitar Corp., 1630 Stewart Street, Santa Monica, Calif. 90406. Vivitar telephoto lenses.

Welt/Safe-Lock, Inc., Hialeah, Fla. Tripods and portable projection tables.

Yashica, Inc., 411 Sette Drive, Paramus, N.J. 07652. Cameras, telephoto lenses.

FEEDERS, GRAIN, AND TEACHING AIDS

Aspects Inc., P.O. Box 23, Bristol, R.I. 02809. Window bird feeders and accessories.

Audubon Workshop, 1501 Paddock Drive, Northbrook, Ill. 60062. Bird feeders, houses, baths, grain and other bird foods.

Bird Friends Society, Essex, Conn. 06426. Bird feeders, houses, baths, grain and other bird foods. Catalog fee.

Bird Seed Savings Day, Division of Nature Center Associates, Inc., 16 Holmes Street, Mystic, Conn. 06355. Bird feeders, houses, baths, grain and other bird foods.

The Bird Tree, 5 Swallow Lane, St. Paul, Minn. 55110. Bird feeders, houses, baths, grain, and other accessories.

The Bower Mfg. Co., Inc., P.O. Box 116, Goshen, Ind. 46526. Bird feeders.

Cornell Laboratory of Ornithology, 159 Sapsucker Woods Road, Ithaca, N.Y. 14850. Bird feeders, houses, selection of bird slides.

Coronet Instructional Media, 65 E. South Water Street, Chicago, Ill. 60601. Bird filmstrip series.

Dialbird, 554 Chestnut Street, Westwood, N.J. 07675. Bird feeders and houses.

Duncraft, Wild Bird Specialists, Penacook, N.H. 03303. Bird feeders, houses, baths, grain and other accessories.

Educational Images, P.O. Box 367, Lyons Falls, N.Y. 13368. Slide sets and filmstrips for teaching purposes.

Encyclopedia Britannica Educational Corporation, 425 N. Michigan Avenue, Chicago, Ill. 60611. Distributor of "Audubon Aids" about birds and wildlife produced by the National Audubon Society.

Gull Lake Environmental Education Project, Kellogg Bird Sanctuary of Michigan State University, Augusta, Mich. 49012. Teaching aids: filmstrips, slides, flashcards, charts, packets.

Hallco Products, 718 First Avenue, N.W., Grand Rapids, Minn. 55744. Bird feeders, houses, baths.

Hummingbird Heaven, 1254 Carmel Drive, Simi, Calif. 93065. Bird feeders, houses, baths.

Hyde Bird Feeder Company, 56 Felton Street, Box 168, Waltham, Mass. 02254. Bird feeders, houses, baths, grain and other bird foods.

Jae Jae Industries, 3141 Oak Street, Santa Ana, Calif. 92707. Birdhouses.

Jamie Wood Birds Ltd., Cross Street, Polegate, Sussex, England. Bird blinds.

Massachusetts Audubon Society Gift Shop, S. Great Road, Lincoln, Mass. 01773. Bird feeders, houses, baths, grain and other bird foods. Catalog fee.

National Audubon Wild Bird Seed, National Audubon Society, P.O. Box 117, Bristol, Ill. 60512. Grain and other bird foods.

Nature House Inc., Purple Martin Junction, Griggsville, Ill. 62340. Books on purple martins, martin houses.

Nelson Manufacturing Company, 3049 12th Street S.W., Cedar Rapids, Iowa 52406. Bird feeders, houses, baths.

Norshore Pets, 6206 S. Route 23, Box 271, Marengo, Ill. 60152. Bird feeders, houses, baths, grain and other bird foods. Catalog fee.

Postmart, Box 473, Avon, Conn. 06001. Water warmer for bird baths.

Tamrac, P.O. Box 4690, N. Hollywood, Calif. 91607. Bird-watcher's belt bag for bird guides.

Welles L. Bishop Co., 1245 E. Main Street, Meriden, Conn. 06450. Bird feeders, houses, grain and other bird foods.

Wild Bird Supplies, 4815 Oak Street, Crystal Lake, Ill. 60014. Bird feeders, houses, baths, grain and other bird foods.

Wildlife Refuge, 8845 W. Bath Road, Laingsburg, Mich. 48848. Bird feeders with squirrel baffles.

B

BIRD BOOK
RETAILERS

ABA Sales (American Birding Association), P.O. Box 4335, Austin, Texas 78765.

Audubon Bookcase, 3890 Stewart Road, Eugene, Oreg. 97402. New books.

Audubon Workshop, 1501 Paddock Drive, Northbrook, Ill. 60062. New books, sound recordings.

Bird Friends Society, Essex, Conn. 06426. New books, sound recordings. Catalog fee.

Birding Book Society, Riverside, N.J. 08370. Discounts on new books.

Bohdan Zaremba, 3 Livermore Place, Cambridge, Mass. 02141. Used and out-of-print books. Catalog fee.

The Book Chest, Inc., 197 Oxford Place, Rockville Centre, N.Y. 15570. Scarce and rare books. Catalog fee.

Books About Birds, P.O. Box 106, Kew Gardens, N.Y. 11415. A newsletter about new, scarce, and rare books. Subscription fee.

Books-On-File, Union City, N.J. 07087. Rare and used books.

Box Windows Book Shop, 128 High Street, Lewes, East Sussex BN7 1XL, England. Used books.

Buteo Books, P.O. Box 481, Vermillion, S. Dak. 57069. New, used, and out-of-print books. Fee for first catalog, subsequent catalogs free. Partial listing free.

The Chickadee, 702 Marshall Street, Houston, Tex. 77006. New books.

Cornell Laboratory of Ornithology, 159 Sapsucker Woods Road, Ithaca, N.Y. 14850. New books, complete selection of sound recordings.

Duncraft, Wild Bird Specialists, Penacook, N.H. 03303. New books, particularly about attracting birds.

Falconiforme Press, Ltd., P.O. Box 4047, Saskatoon S7K 3T1, Saskatchewan. Books about falconry, hawks, and owls.

Heffers Booksellers, 20 Trinity Street, Cambridge CB2 3NG, England. New books.

John Johnson, R.F.D. 2, North Bennington, Vt. 05257. Used books.

Patricia Ledlie, Bookseller, Box 46, Buckfield, Maine 04220. Used and out-of-print books. Catalog fee.

Lloyd Imig Bird Books, 310 Maria Drive, Wausau, Wis. 54401. New and out-of-print books. Catalog fee (refundable with first order).

Los Angeles Audubon Society, Audubon House, 7377 Santa Monica Boulevard, Los Angeles, Calif. 90046. New North American and international bird books, lists, and records.

Lucas Book Co., 2430 Bancroft Way, Berkeley, Calif. 94704. Used books.

Massachusetts Audubon Society Gift Shop, S. Great Road, Lincoln, Mass. 01773. New books, sound recordings. Catalog fee.

National Audubon Society, The Booknest, Western Education Center, 376 Greenwood Beach Road, Tiburon, Calif. 94920. New books, sound recordings.

National Audubon Society, Service Department, 950 Third Avenue, New York, N.Y. 10022. New books, sound recordings.

The Natural History Bookcase, 3136 Brophy Drive, Sacramento, Calif. 95821. New and used books.

Nature Canada Bookshop, 75 Albert Street, Ottawa K1P 6G1, Ontario. New books.

Petersen Book Company, P.O. Box 966, Davenport, Iowa 52806. New and used books.

Rudolph Wm. Sabbot, Natural History Books, 5239 Tendilla Avenue, Woodland Hills, Calif. 91364. Out-of-print and imported books.

Tolliver's Books, 1634-AA Stearns Drive, Los Angeles, Calif. 90035. New and out-of-print books.

Watkins Natural History Books, Rebecca and Larry C. Watkins, R.D. #1, Belden Corners Road, Dolgeville, N.Y. 13329. Used books.

Whelden & Wesley Ltd., Lytton Lodge, Codicote, Hitchin, Hertfordshire SG4 8TE, England. Used books.

Wild Bird Supplies, 4815 Oak Street, Crystal Lake, Ill. 60014. New books.

William R. Hecht, Box 67, Scottsdale, Ariz. 85252. Used books.

C

BIRD
AND BIRD-WATCHING
PUBLICATIONS
FROM NATURAL
RESOURCE AGENCIES

FREE PUBLICATIONS

U.S. Department of the Interior

U.S. FISH AND WILDLIFE SERVICE
Free single copies of the following publications may be obtained from the Publications Unit, U.S. Fish and Wildlife Service, Department of the Interior, Washington, D.C. 20240.

Conservation Notes: Four- to eight-page illustrated discussions of conservation topics. Bird titles include the following: "Birds" (Conservation Note No. 1), "Protecting Our Endangered Birds" (Conservation Note No. 3), "America's Upland Game Birds" (Conservation Note No. 4), "Bird Banding: The How & Whys" (Conservation Note No. 5), "Migration of Birds" (Conservation Note No. 8), and "The Bald Eagle" (Conservation Note No. 20).

Wildlife Biologue Series: The Biologue series answers many commonly asked questions about specific species and conservation-related topics. Each Biologue presents two pages of biologically based information about fish and wildlife species or environmental issues. In addition to the following list of bird titles, the series also contains topics about mammals, fish, and reptiles. Write the Fish and Wildlife Service Publications Unit for a complete list. **Bird Titles:** Accipiters: Cooper's Hawk, Goshawk, Sharp-shinned Hawk; Blackbirds; Masked Bobwhite; Buteos: Broad-winged Hawk, Ferruginous Hawk, Red-shouldered Hawk, Rough-legged Hawk, Swainson's Hawk; California Clapper Rail; California Condor; Sandhill Crane; Dodo; Canvasback Duck; Eagles: Bald Eagle, Golden Eagle; Snowy Egret; Falcons: Peregrine, Kestrel, Merlin, Pigeon Hawk, Sparrow Hawk, Prairie Falcon; Aleutian Canada Goose; Canada Goose; Marsh Hawk and Osprey; Red-tailed Hawk; Great Blue Heron; Hummingbirds; Great Horned Owl; Snowy Owl; Owls: Barn Owl, Long-eared, Screech, Short-eared, Barred, Spotted; Owls: Great Horned, Saw-whet, Pygmy, Snowy, and Burrowing Owls; Brown Pelican; Dusky Seaside Sparrow; Trumpeter Swans; California Least Tern; Wild Turkeys; Turkey Vulture; Woodcock; American Ivory-billed Woodpecker. **Conservation Titles:** Build a Bluebird House; Endangered Species; What Your Club Can Do for Habitat; You Asked About Hunting?; Purple Martin House Design; Westlands; A Modern Day Threat to Wildlife; The Fish and Wildlife Service; What Can I Do for Wildlife?; Winter Feeders; Laws Protecting Birds.

Miscellaneous Publications: Attracting and Feeding Birds (Conservation Bulletin No. 1); Bald Eagles in Alaska; Extinct Wildlife; Facts About Federal Wildlife Laws; Fish, Wildlife, and People; Puddle Ducks; Something About Hawks; Tricks to Control Birds; Whooping Cranes.

National Wildlife Refuge Publications: "Directory of National Wildlife Refuges." Arranged by states, the directory indicates the county in which each refuge is located, date of establishment, acres, and principal species protected. Also includes a one-page map.

"Address List of Refuge Managers." A complete list of refuge managers, their addresses, and phone numbers. Also includes a one-page map.

"A Visitor Guide to the National Wildlife Refuges." A U.S. map showing location of all national wildlife refuges.

"Refuge Leaflets" and "Bird Lists" are available for the following U.S. national wildlife refuges and regions:

	Refuge Leaflet	Bird List		Refuge Leaflet	Bird List
National Wildlife Regions			*Alabama*		
Region 1	X		Choctaw	X	X
Region 2	X		Eufaula	X	X
Region 3	X		Wheeler	X	X
Region 4	X				
Region 5	X				
Region 6	X				

	Refuge Leaflet	Bird List		Refuge Leaflet	Bird List
Alaska			San Joaquin	X	
Alaskan Refuges	X		San Luis	X	X
Aleutian Islands	X	X	San Pablo	X	
Arctic	X	X	Sutter	X	X
Bering Sea		X	Tule Lake	X	X
Clarence Rhode	X	X			
Izembek	X	X			
Kenai	X	X	**Colorado**		
Kodiak	X	X	Alamosa		X
			Arapaho		X
Arizona			Browns Park	X	
Cabeza Prieta	X	X	Monte Vista	X	X
Cibola	X	X			
Havasu	X	X			
Imperial	X	X	**Delaware**		
Kofa	X	X	Bombay Hook	X	X
			Prime Hook	X	
Arkansas					
Big Lake	X	X	**Florida**		
Holla Bend	X	X	Chassahowitzka	X	X
Wapanocca	X	X	Florida Keys	X	X
White River	X	X	Great White Heron	X	X
			J. N. "Ding" Darling	X	X
California			Key West	X	X
Cibola	X	X	Lake Woodruff	X	X
Clear Lake	X	X	Loxahatchee	X	X
Colusa	X	X	Merritt Island	X	X
Delevan	X	X	National Key Deer	X	X
Farallon	X		Okefenokee	X	X
Havasu	X	X	Pelican Island	X	X
Imperial	X	X	St. Marks	X	X
Kern	X	X	St. Vincent	X	
Kesterson	X				
Klamath Basin	X	X	**Georgia**		
Lower Klamath	X	X	Blackbeard Island	X	X
Merced	X	X	Eufaula	X	X
Modoc	X	X	Harris Neck	X	X
Pixley	X	X	Okefenokee	X	X
Sacramento	X	X	Piedmont	X	X
Salton Sea	X	X	Savannah	X	X
San Francisco Bay		X	Wassaw		

	Refuge Leaflet	Bird List		Refuge Leaflet	Bird List
Hawaii			*Maryland*		
Hawaiian Islands	X	X	Blackwater	X	X
			Chincoteague	X	X
Idaho			Eastern Neck	X	X
Camas	X	X			
Deer Flat	X	X	*Massachusetts*		
Grays Lake	X	X	Great Meadows	X	X
Kootenai	X	X	Monomy	X	X
Minidoka	X	X	Oxbow	X	
			Parker River	X	X
Illinois					
Chautauqua	X	X	*Michigan*		
Crab Orchard	X	X	Seney	X	X
Mark Twain	X	X	Shiawssee	X	X
Upper Mississippi	X	X			
			Minnesota		
Indiana			Agassiz	X	X
Muscatatuck	X		Big Stone	X	X
			Rice Lake	X	X
Iowa			Sherburne	X	X
De Soto	X	X	Tamarac	X	X
Mark Twain	X	X	Upper Mississippi	X	X
Union Slough	X	X			
Upper Mississippi	X	X	*Mississippi*		
			Noxubee	X	X
Kansas			Yazoo	X	X
Flint Hills	X	X			
Kirwin	X	X	*Missouri*		
Quivira	X	X	Mingo	X	X
			Squaw Creek	X	X
Kentucky			Swan Lake	X	X
Reelfoot	X	X			
			Montana		
Louisiana			Benton Lake	X	X
Breton	X		Bowdoin	X	X
Delta	X	X	Charles M. Russell	X	X
Lacassine	X	X	Medicine Lake	X	X
Sabine	X	X	National Bison	X	X
			Ninepipe	X	X
Maine			Pablo	X	X
Moosehorn	X	X	Ravalli		X
Rachel Carson	X		Red Rock Lakes	X	X

	Refuge Leaflet	Bird List		Refuge Leaflet	Bird List
Nebraska			Des Lacs	X	X
Crescent Lake	X	X	J. Clark Salyer	X	X
DeSoto	X	X	Lake Ilo	X	
Fort Niobrara	X	X	Long Lake	X	X
Valentine	X	X	Lost Wood	X	X
			Slade	X	X
Nevada			Souris Loop		X
Anaho Island	X	X	Sully Hill	X	X
Charles Sheldon	X	X	Tewaukon	X	
Desert	X	X	Upper Souris	X	X
Pahranagat		X			
Ruby Lake	X	X	*Ohio*		
Sheldon	X	X	Ottawa	X	X
Stillwater	X	X			
			Oklahoma		
New Jersey			Salt Plains	X	X
Brigantine	X	X	Sequoyah	X	X
Great Swamp	X	X	Tishomingo	X	X
			Washita	X	X
New Mexico			Wichita Mts.	X	X
Bitter Lake	X	X			
Bosque del Apache	X	X	*Oregon*		
Las Vegas	X	X	Ankeny	X	X
Maxwell	X	X	Baskett Slough	X	X
San Andres	X	X	Charles Sheldon	X	X
			Cold Springs	X	X
New York			Columbian White-		
Iroquis	X	X	tailed Deer	X	
Montezuma	X	X	Deer Flat	X	X
Morton	X	X	Hart Mountain	X	X
Target Rock	X	X	Klamath Basin	X	X
			Lewis & Clark	X	
North Carolina			Lower Klamath	X	X
Mackay Island		X	Malheur	X	X
Mattamuskeet	X	X	McKay Creek	X	X
Pea Island	X	X	Sheldon	X	X
Pungo	X		Umatilla		X
Swanquarter		X	Upper Klamath	X	X
			W. L. Finley	X	X
North Dakota					
Arrowwood	X	X			
Audubon	X	X			

	Refuge Leaflet	Bird List		Refuge Leaflet	Bird List
Pennsylvania			**Vermont**		
Erie	X	X	Missisquoi	X	X
South Carolina			**Virginia**		
Cape Romain	X	X	Back Bay	X	X
Carolina Sandhills	X	X	Chincoteague	X	X
Santee	X	X	Mackay Island		X
Savannah	X	X	Mason Neck		X
			Presquile	X	X
South Dakota			**Washington**		
Lacreek	X	X	Columbia	X	X
Lake Andes	X	X	Columbian White-tailed Deer	X	
Sand Lakes	X	X	Dungeness	X	X
Waubay	X	X	McNary	X	X
			Nisqually	X	
Tennessee			Puget Sound	X	
Cross Creeks	X	X	Ridgefield	X	X
Hatchie	X	X	Saddle Mountain	X	
Lake Isom	X	X	Turnbull	X	X
Reelfoot	X	X	Umatilla		X
Tennessee	X	X	Washington Coast	X	
			Willpa	X	X
Texas			**Wisconsin**		
Anahuac	X	X	Horican	X	X
Aransas	X	X	Necedah	X	X
Attwater Prairie Chicken	X	X	Upper Mississippi	X	X
Brazoria		X			
Buffalo Lake	X	X	**Wyoming**		
Hagerman	X	X	Hutton Lake		X
Laguna Atascosa	X	X	National Elk	X	X
Muleshoe	X	X	Seedskadee	X	X
San Bernard		X			
Santa Ana	X	X			
Utah					
Bear River	X	X			
Fish Springs	X	X			
Ouray	X	X			

NATIONAL PARK SERVICE
Single copies of the following publications may be obtained by writing to:
National Park Service, Washington, D.C. 20240.
"Index to the National Park System and Related Areas." A complete list of
   addresses for national parks and related areas with a brief description of
   significant features and total acreage.
"Map of the National Park System." U.S. map showing location of national
   parks, seashores, monuments, and other facilities within the National Park
   System.

U.S. Department of Agriculture

PUBLICATIONS DIVISION
Single copies of the following publications may be obtained free by writing to:
Publications Division, Office of Governmental and Public Affairs, U.S.
Department of Agriculture, Washington, D.C. 20250.
"More Wildlife Through Soil and Water Conservation." 1971. Stock No.
   AB-175.
"Building a Pond." 1973. Stock No. F-2256.
"Making Land Produce Useful Wildlife." 1969. Stock No. F-2035.
"Ponds and Marshes for Wild Ducks on Farms and Ranches in the Northern
   Plains." 1968. Stock No. F-2234.
"Autumn Olive for Wildlife and Other Conservation Uses." 1972. Stock No.
   L-458.
"Russian-olive for Wildlife and Other Conservation Uses." 1971. Stock No.
   L-517.
"Wildlife for Tomorrow." 1972. Stock No. PA-989.

FOREST SERVICE
Single copies of the following publications may be obtained free by writing to:
Forest Service, U.S. Department of Agriculture, P.O. Box 2417, Washington,
D.C. 20013.
"Discover Birding in the National Forests." A general brochure reviewing some
   of the bird-watching opportunities available in the national forests and
   grasslands.
"Field Offices of the Forest Service." U.S. map with addresses for all U.S.
   national forests and grasslands. Maps of specific national forests may be
   obtained by writing to individual national forests and grasslands.

Environment Canada

CANADIAN WILDLIFE SERVICE
Individuals may order, without cost, up to ten different publications; schools,
libraries, natural history clubs, sportsman's clubs, and other organizations may
order one complete set of pamphlets provided these requests are written on the
letterhead of the organization. Mail requests for publications to: Distribution
Section, Canadian Wildlife Service, Environment Canada, Ottawa K1A OE7,
Ontario.

Hinterland Who's Who Series: Four to six-page, illustrated, $8\frac{1}{2} \times 11$-inch pamphlets describing the life histories of birds and mammals and directed toward students and adults with an interest in wildlife. **Bird Titles:** American Robin; American Woodcock; Black-capped Chickadee; Black Duck; Blue Jay; Bufflehead; Canada Goose; Canvasback; Downy Woodpecker; Evening Grosbeak; Gannet; Gray Jay; Great Blue Heron; Great Horned Owl; Greater Snow Goose; Hawks (accipiters); Herring Gull; Killdeer; Loon; Mallard; Mountain Bluebird; Peregrine Falcon; Ptarmigans; Red-breasted Nuthatch; Redhead; Ring-billed Gull; Ring-necked Pheasant; Ruby-throated Hummingbird; Ruffed Grouse; Snowy Owl; Trumpeter Swan; Whippoorwill; Whistling Swan; Wood Duck; Whooping Crane.

Attracting Birds Series: Plans and guidelines for constructing bird houses and feeders in similar format to the *Hinterland Who's Who Series:* **Titles:** Bird Feeders; Nest Boxes for Birds.

Wildlife Interpretation Centres: Colorful brochures describing the unique wildlife attractions and travel directions to the CWS Wildlife Centres.

SOIL CONSERVATION SERVICE

Single copies of the following publications may be obtained by contacting your local Soil Conservation Service office listed in the telephone directory under U.S. Government: Department of Agriculture. Free copies may also be obtained by writing to: Soil Conservation Service, Information Division Room 0054, Washington, D.C. 20013.

"Invite Birds to Your Home: Conservation Plantings for the Northeast."
"Invite Birds to Your Home: Conservation Plantings for the Southeast."
"Invite Birds to Your Home: Conservation Plantings for the Midwest."
"Invite Birds to Your Home: Conservation Plantings for the Northwest."
"Conservation Plants for the Northeast."

PUBLICATIONS FOR SALE

U.S. Government Printing Office

The following publications are for sale from the Superintendent of Documents. To order and obtain a current price list write to: Superintendent of Documents, U.S. Government Printing Office, Washington, D.C. 20402.

"Birds in Our Lives." Shows the influence of birds on arts, literature, design; describes search for knowledge of how birds migrate; covers birds and pesticides, hunting, and introduction of exotic species. Written by 61 authors. 80 wash drawings and 372 photos. Cloth. 1974. 576 pages. Stock No. 2410-00001.

"California Condor, 1966–1976: A Look at Its Past and Future." The result of an extensive literature search and 900 field days, this book reviews and analyzes the population and distribution of the California condor, discusses its prospects for the future, and summarizes preservation plans. 1978. 136 pages. Illustrated. Stock No. 024-010-00453-3.

"Cavity-Nesting Birds of North American Forests." This lively booklet provides full-color illustrations of each of the 85 species of North American birds that nest in cavities. It also summarizes published and observed data on the favored nesting habitats and dietary requirements of each of the species. Small range maps indicate where the given species are likely to be located in different seasons. 1977. 112 pages. Illustrated. Stock No. 001-000-3726-9.

"Conservation, Migratory Birds, Agreement Between the United States of America and the Union of Soviet Socialist Republics." Stock No. 044-000-91928-1.

"Diving Ducks." Describes life habits and characteristics of diving ducks. 1971, reprinted 1974. 16 pages. Illustrated. Stock No. 024-010-00254-9.

"Duck Stamp Data." Presents the history and purpose of the Migratory Bird Hunting and Conservation Stamp (the "Duck Stamp"). Printed in loose-leaf format to fit standard, 3-ring binders so that it can be brought up to date annually. Each year updated pages will be stocked by GPO and their availability announced in GPO's periodic list of federal government publications. Alternatively, "Duck Stamp Data" can be ordered on a subscription basis with automatic updates each year. 1978. 52 pages. Illustrated. Stock No. 024-010-00455-0.

"Ducks at a Distance: A Waterfowl Identification Guide." Revised 1978. 52 pages. Illustrated. Stock No. 024-010-00442-8.

"The Fairest One of All." Describes over twenty-two endangered, rare, or unique wildlife species that reside in the eastern United States. 32 pages. Illustrated. Stock No. 001-001-00311-5.

"Fifty Birds of Town and City." This book has pictures of and describes fifty birds commonly seen in U.S. cities, with full-color illustrations that enable you easily to identify each bird when you see it. Cloth and paper. 1974. 50 pages. Illustrated. Stock No. 024-010-00382-1.

"Homes for Birds." Conservation Bulletin No. 14. This pamphlet reviews some of the principles of birdhouse design, the materials necessary to build the homes, and their location around the dwelling. Specific likes and dislikes of various species of songbirds are reviewed. Revised 1974. 18 pages. Illustrated. Stock No. 024-000-00050-8.

"Migration of Birds." Circular 16. This booklet answers some of the most prevalent questions on bird migration, points out the main routes for migration in North America, and discusses techniques for studying migration. Revised edition 1979. 119 pages. Stock No. 024-010-00484-3.

"Nest Boxes for Wood Ducks." 1976. 13 pages. Illustrated. Stock No. 024-010-00415-1.

"Ponds and Marshes for Wild Ducks on Farms and Ranches in the Northern Plains." 1970. 16 pages. Illustrated. Stock No. 001-000-00087-0.

"Planning for Wildlife in Cities and Suburbs." Describes how to create and maintain better environments for both wildlife and people in urban and suburban areas. Many of the general approaches discussed in the booklet can be applied in the backyards of private homes. 1978. 64 pages. Stock No. 024-010-00471-1.

"Symbol of Our Nation: The Bald Eagle." An 11½ × 15-inch color reproduction of the bald eagle, mounted on heavy paper and suitable for framing. Includes a white border; a brief description of the precarious situation facing the survival of the nation's symbol is also provided. Stock No. 2410-00002.

"To Have and to Hold: Alaska's Migratory Birds." This colorful pamphlet describes Alaska's key role in producing migratory waterfowl, most of which are enjoyed by people throughout the rest of the United States. It identifies the vital habitat that must be preserved if these wild flocks are to survive. 1973. 14 pages. Illustrated. Stock No. 2410-0279.

"Visit a Lesser-used Park." Guide listing and briefly describing the less crowded U.S. national parks. Stock No. 024-005-0589-7.

"Waterfowl Tomorrow." For all outdoorsmen and women, hunters, naturalists, and conservationists. Story of migratory waterfowl on the North American continent, popularly written by experts from the United States, Canada, and Mexico. 1964, reprinted 1975. 770 pages. Illustrated. Cloth. Stock No. 024-010-00013-9.

Wildlife Portrait Series

No. 1. "Wildlife Portraits." Printed in full color, this set of ten 14 × 17-inch wildlife reproductions features color photographs of birds and other wildlife species. 1969. Stock No. 024-010-00190-9.

No. 3. "Songbirds." Ten attractive 14 × 17-inch posters featuring common songbirds. The set includes a pamphlet reviewing field marks and behavior of each species. 1974. Stock No. 024-010-00388-0.

No. 4. "A Host of Seabirds—Alaska." Six 16 × 22-inch prints by artist Bob Hines featuring the spectacular multitude of puffins, auklets, and shearwaters that populate Alaska's rich coastal waters. The set includes an informative pamphlet with background information for each poster. 1979. Stock No. 024-010-00530-1.

Government of Canada Publications

The following publications are for sale by the Canadian Publishing Centre. For orders and current prices write to: Publishing Centre, Printing and Publishing, Supply and Services Canada, Ottawa K1A OS9, Ontario.

"The Autumn Birds of Point Pelee National Park." List of autumn birds (from August 16 to November 15). Special ornithological events of the autumn period. 1973. 33 pages. Paperback. Stock No. R63-7273.

"The Spring Birds of Point Pelee National Park." List of spring birds (from March 1 to June 10). Special ornithological events of the spring period. 1973. 32 pages. Paperback. Stock No. R63-2872.

"The Summer Birds of Point Pelee National Park." List of summer birds (from June 11 to August 15). Special ornithological events of the summer period. 1973. 28 pages. Paperback. Stock No. R63-7172.

"The Winter Birds of Point Pelee National Park." List of winter birds (from November 16 to February 29). Special ornithological events of the winter period. 1973. 25 pages. Paperback. Stock No. R63-7072.

"Birds of Canada." 518 species described as to measurements, field marks, habitat, nesting, range status in Canada, English, French, and scientific names. Most of the species depicted in color by artist J. A. Crosby. Also ink drawings of special features of several species. Maps showing distribution. Glossary. Index. 1974. 428 pages. Cloth. Stock No. NM93-203.

"Ducks at a Distance." A waterfowl identification guide with numerous color and black-and-white illustrations. 1973. 20 pages. Paperback. Stock No. CW66-3665.

"Sea-birds of Bonaventure Island." Gannet, double-crested cormorant, great black-backed gull, herring gull, black guillemot, black-legged kittiwake, razorbill, common murre, common puffin, Leach's petrel. Description, nesting, and color photograph for each. 1973. 20 pages. Paperback. Stock No. CW66-4173.

"Some Canadian Birds." Short illustrated account of some thirty common species widely distributed in Canada. Ink drawings. 1972. 44 pages. Paperback. Stock No. NM92-566.

"Ornamental Shrubs for Canada." Guide to buying, growing, and caring for ornamental shrubs. Description of some 150 species. Bibliography. Index. Eighty illustrations, of which 30 are in color. Fold-out maps showing plant hardiness zones. 1974. 187 pages. Paperback. Stock No. A53-1286.

Photo by Henry A. Harding, York Harbor, Maine

INDEX

Illustrations are noted in *italics*.

age of birds: color and, 5; size and, *7*

anatomy. *See* bird anatomy

anting, *42*

appeasement displays, *52*

approaches. *See* closing the distance

attraction techniques: pishing, 13; squeaking, 13, *14*. *See also* bird blinds

Audubon Bird Call, 13, *14*

Audubon Christmas Bird Count, 72

bathing, *41*

behavior, characteristic, 2, 4. *See also* behavior watching

behavior watching: bird senses *vs.* human senses, 40; body care behaviors, 40–44; and feeding strategies, 44–51; guidelines for, 39–40; social displays, *52–57*. *See also* behavior-watching activities

behavior-watching activities, 57–62; ethogram building, 57–59; and publication of observations, 62

bill wiping, *42*

binocular selection, 24–33; and alignment, 29, *30*, 33; and center focus *vs.* individual focus design, *30*, 31; for eyeglass wearers, 32; and field of view, 27, *28*, 29; and light-gathering capacity, 25–27; and mini binoculars, 32; and power, 24–25; and resolution, 29; and roof prism binoculars, *31*; tests for, 32–33. *See also* binoculars

binoculars: behavior watching and, 61; care of, 34–35; cleaning, 34; and coated lenses, 27; comparisons of exit pupils, 25, *26*, *27*; falling into water, 35; finding birds with, 17; focusing, 33–34; lenses and prisms of, *25*; protecting, 34–35; reflection at each surface of, *28*; repairs, 29, 31; use on field walk, 23; use near windows, 22. *See also* binocular selection

bird anatomy, *73–78*; flight postures, *78*

bird blinds, 97–104; aluminum cube, *103*; bird feeder as, *98*; with convenience perch, *101*; atop homemade scaffold, *100*; ice-fishing huts as, *100*; portable, designs for, 102, *103–4*; and wary species, 98; wooden-frame, *104*

bird counts. *See* counting birds

bird feeder, as bird blind, *98*

bird leg, perched, *75*

bird sounds. *See* calls; recordings; songs

bird walks, tips for, 21–23

319

bird-watching: and locating birds, 17–20; with small reflector telescope, 37–38; and spotting scopes, 35, 37; and tips for bird walks, 21–23. *See also* behavior watching; identification of birds

Birder's Field Notebook, The, 63

blocking, *71*; practice patterns for, *72*

boats. *See* nautical birding

body care behaviors, *40–44*; in ethogram, 58–59; ritualized, 52

breeding season, pishing in, 13

brood patch, 55

brooding, *55*

buddy binoculars, *20*

calls, *53*; pishing and squeaking, 13–14. *See also* songs

camera selection, 83–87; telephoto lenses, 84–87

cardioid microphones, 114–15

caution displays, *52*

censuses and surveys of birds, 73

center focus binoculars, *30*, 31

chase response, recordings causing, 16

chip calls, 13

chiseling, *46*

"clock" technique, 17, *18, 19*, 22

closing the distance, 13–17; and mobbing behavior, *15, 16*; and pishing, 13, 14; and recorded songs, 16; and squeaking, 13, *14*; and treetop birding, 17

clothing, for bird walks, 22

color, bird identification by, 5–6

color patterns. *See* field marks

condenser microphones, 114

copulation, *54*

costs. *See* prices

counting birds, 70–73; and blocking, *71, 72*

courtship displays, 52, *54*, 58

courtship feeding, *54*

dabbling, *50*

displacement activities, 52

distraction displays, *57*

distress call. *See* pishing

diving, *49*

dusting, *41*

dynamic microphones, 114

eggs: hatching, bird blinds and, 97–98; turning, 55

endangered species. *See* rare birds

ethogram building, 57–59; clocking time budget in, 59, *61*; identifying behavioral sequence in, 59

etiquette: in field, 22; for photographers, 105

exit pupils of binocular: blocked and unblocked, *27*; brightness and, 25, *26*; tests for, 32

extension tubes, 85

eyeglass wearers, binoculars for, 32

families, 1; characteristic behaviors of, 2, 4; characteristic shapes of, *2*; knowledge of, 13

feather ruffling and fluffing, *43*

feather settling, *43*

feathers, secondary, *77*

feeding strategies, 44–51; characteristic of families, 2, 4. *See also* courtship feeding

field equipment, insurance for, 38. *See also* binoculars; spotting scopes, tripods

field identification report, form for, 63

field marks, 4, *5*, 23

field note taking: and *The Birder's Field Notebook*, 63; and checklists, 62; equipment for, 69–70; notebooks for, 62–63. *See also* field notes; Grinnell Field Note System

field notes: and data punch cards, 70; preservation of, 70; sample of, *60*

filing system for bird slides, 106–8

flash equipment, 87–89; use of, 97

flight patterns, *78*; characteristic among families, 2, *3*

flocking, *53*; counting and, 70, *71, 72*, 73

flying birds, photographing, 94

food carrying, *56*

foot raking and stamping, *49*

freezing, *8, 56*

gleaning. *See* hover gleaning; perch gleaning; sally gleaning

grazing, *50*

Grinnell Field Note System, 64–70

habitat, *8*, 9, *10*; behavior and, 39; in photographs, *91*
hawking, *45*
hearing sense of birds, 40
hover gleaning, *46*

identification of birds, 1–13; behaviors and, 2, 4; characteristic flight patterns and, 2, *3*; color and, 5–6; field marks and, 4, *5*; habitat and, 8–9, *10*; knowledge of families and, 13; range and, 10–11; season and, *12*, 13; shapes and, 2; similar postures among species and, 2, *3*; size comparisons and, *6*, 7; songs and, 7–8; species abundance and, 11
incubation, *55*
individual focus binoculars, *30*, 31
insurance for field equipment, 38

journal: abbreviations, 65; writing guidelines, 69–70

leaf tossing, *46*
lenses of binoculars, *25*, 26–27; tests for, 32. *See also* telephoto lenses
light table for evaluating slides, *107*
lighting: and color of birds, 5; and counting birds, 73; and mini binoculars, 32; for photographs, *96*; and size of birds, 6
locating birds, 17–20; and buddy binocular, *20*; "clock" technique, 17, *18*, 19; need for specific descriptions of locations, 17

macrozoom lenses, 86
microphones, 114–15; fields of sensitivity for, *114*
migration: and habitat, *9*; and range maps, 10, *11*
mini binoculars, 32
mirror, to locate birds, 19
mobbing behavior, *15*, *16*, *53*; causing, 15, *16*
molting, 5; and eating of feathers, 39
mounting, *54*

nautical birding: and binoculars, 26; "clock" technique for, *18*, 19
nectar hovering, *51*
nest building, *55*
nest cleaning, *56*
nests, and bird blinds, 97–98
noise reduction, recording and, 115

oiling, *41*
omnidirectional microphones, 114

packing, *51*
panting, *42*
perch gleaning, *45*
photographic equipment selection, 83–90. *See also* cameras; flash equipment
photographic problems: with angle of view, *92*; blurred images, *94*; distracting backgrounds, *92*, *93*; extraneous habitat, *91*; with lighting, *96*, 97; neglecting background detail, *93*; over- and underexposure, *95*; slanting horizons, *93*
photography: advantages of, 83; common birds as subjects for, *99*; etiquette for, 105; filing system for slides, 106–8; film processing and printing, 90; film selection, 90; home-study course in, 106; and recordings, 83, 115; and use of bird blinds, *see* bird blinds. *See also* photographic equipment selection; photographic problems; slide programs
piracy, *49*
pishing, 13, 14
plucking, *51*
plunging, *48*
postures, 2, *3*
pounding, *47*
predators: and distraction displays, *57*; and mobbing behavior, 15–16
preening, *40*
prices: of binoculars, 24, 29, 31, 32, 33; of insuring field equipment, 38; of small telescopes, 38; of tape recorders, 113
probing, *46*
pruning, *51*

range maps, 10, *11*

ranges, 10–11

rare birds: on field walks, 22; nesting, tape recordings and, 16

recordings, 113–15; and attracting birds, 16; of bird songs to help recognize them, 8; equipment selection for, 113–15; and microphones, 114–15; of screech owl, mobbing reaction and, 15–16; and slide programs, 83. *See also* tape recorders

reel-to-reel recorders, 113–14

regurgitation, of nondigestible materials, 39–40

rodent running, *57*

roof prism binoculars, *31*

sally gleaning, *45*

scavenging, *50*

scratching, *44*

season: and bird identification, *12*, 13; and color, 5

senses, of birds and humans compared, 40

sex: color and, 5; size and, 7

shapes, characteristic, *2*

shell smashing, *48*

shotgun microphones, 114

sifting, *50*

silhouettes: flight, *3*; postures, *3*; shapes, *2*; for size comparisons, *6*

size: comparisons, *6*, 7; of young birds, *7*

sketching birds, 73–81; examples of, *79*, *80*

sleeping, *44*

slide programs, 109–13; audiences for, 109; resource materials for, 112–13; techniques for, 110–12

slides: cabinets for, *109*; labs for duplicating, 112–13; and light table, *107*; page from catalog of, *108*

social displays, *52–57*; in ethogram, 58–59

songs, *54*; and identification, 7–8. *See also* recordings

species, 1; characteristic behaviors of, 4; estimated number of, 106; recognition among birds of, 9; seasonal differences in, 12–13; similar postures among, 2, *3*; sizes among, 7

Species Accounts notebook, 64–65; field sketches in, 81; guidelines for entries in, 69–70; sample pages from, *66*, *67*

spotting scope(s), 35, *36*; selection of, 35, 37–38; small reflector telescope, *37*, 38

squeaking, 13, *14*

stalking, *49*

stooping, *47*

stretching, *43*

surface feeding, *48*

sweeping, *47*

tail wagging, *4*

tape recorders: and field notes, 64; microcassette, 63; selection of, 113–14. *See also* recordings

teleconverters, 86

telephoto lenses: all-glass *vs.* mirror, 86; automatic *vs.* present, 86; close-focusing capability of, 84; and depth of field, 85; extension tubes for, 85; and lens speed, 85; magnifying power of, 84; and shutter speeds, 85; supports for, 87, *88*; and teleconverters, 86; zoom, 86

telescopes. *See* spotting scopes

temperature: and binocular alignment, 29; measurement of, *64*; and size comparisons, 6

threat displays, *52*

time budget for bird activity, 59, *61*

treetop birding, 17

tripod(s): flash mounted on, 88; selection of, 38

ultradirectional microphones. *See* shotgun microphones

wind speed, measurement of, *64*

windows, as bird blinds, 97

wings, *76*

zoom lenses, 35, 86, *87*; selection of, 38